PROCEEDINGS OF SYMPOSIA IN APPLIED MATHEMATICS

VOLUME 1   NON-LINEAR PROBLEMS IN MECHANICS OF CONTINUA
Edited by E. Reissner (Brown University, August 1947)

VOLUME 2   ELECTROMAGNETIC THEORY
Edited by A. H. Taub (Massachusetts Institute of Technology, July 1948)

VOLUME 3   ELASTICITY
Edited by R. V. Churchill (University of Michigan, June 1949)

VOLUME 4   FLUID DYNAMICS
Edited by M. H. Martin (University of Maryland, June 1951)

VOLUME 5   WAVE MOTION AND VIBRATION THEORY
Edited by A. E. Heins (Carnegie Institute of Technology, June 1952)

VOLUME 6   NUMERICAL ANALYSIS
Edited by J. H. Curtiss (Santa Monica City College, August 1953)

VOLUME 7   APPLIED PROBABILITY
Edited by L. A. MacColl (Polytechnic Institute of Brooklyn, April 1955)

VOLUME 8   CALCULUS OF VARIATIONS AND ITS APPLICATIONS
Edited by L. M. Graves (University of Chicago, April 1956)

VOLUME 9   ORBIT THEORY
Edited by G. Birkhoff and R. E. Langer (New York University, April 1957)

VOLUME 10   COMBINATORIAL ANALYSIS
Edited by R. Bellman and M. Hall, Jr. (Columbia University, April 1958)

VOLUME 11   NUCLEAR REACTOR THEORY
Edited by G. Birkhoff and E. P. Wigner (New York City, April 1959)

VOLUME 12   STRUCTURE OF LANGUAGE AND ITS MATHEMATICAL ASPECTS
Edited by R. Jakobson (New York City, April 1960)

VOLUME 13   HYDRODYNAMIC INSTABILITY
Edited by R. Bellman, G. Birkhoff, C. C. Lin (New York City, April 1960)

VOLUME 14   MATHEMATICAL PROBLEMS IN THE BIOLOGICAL SCIENCES
Edited by R. Bellman (New York City, April 1961)

VOLUME 15   EXPERIMENTAL ARITHMETIC, HIGH SPEED COMPUTING, AND MATHEMATICS
Edited by N. C. Metropolis, A. H. Taub, J. Todd, C. B. Tompkins (Atlantic City and Chicago, April 1962)

VOLUME 16   STOCHASTIC PROCESSES IN MATHEMATICAL PHYSICS AND ENGINEERING
Edited by R. Bellman (New York City, April 1963)

VOLUME 17   APPLICATIONS OF NONLINEAR PARTIAL DIFFERENTIAL EQUATIONS IN MATHEMATICAL PHYSICS
Edited by R. Finn (New York City, April 1964)
VOLUME 18   MAGNETO-FLUID AND PLASMA DYNAMICS
Edited by H. Grad (New York City, April 1965)
VOLUME 19   MATHEMATICAL ASPECTS OF COMPUTER SCIENCE
Edited by J. T. Schwartz (New York City, April 1966)
VOLUME 20   THE INFLUENCE OF COMPUTING ON MATHEMATICAL RESEARCH AND EDUCATION
Edited by J. P. LaSalle (University of Montana, August 1973)

AMS SHORT COURSE LECTURE NOTES
Introductory Survey Lectures

VOLUME 21   MATHEMATICAL ASPECTS OF PRODUCTION AND DISTRIBUTION OF ENERGY
Edited by P. D. Lax (San Antonio, Texas, January 1976)
VOLUME 22   NUMERICAL ANALYSIS
Edited by G. H. Golub and J. Oliger (Atlanta, Georgia, January 1978)
VOLUME 23   MODERN STATISTICS: METHODS AND APPLICATIONS
Edited by R. V. Hogg (San Antonio, Texas, January 1980)
VOLUME 24   GAME THEORY AND ITS APPLICATIONS
Edited by W. F. Lucas (Biloxi, Mississippi, January 1979)
VOLUME 25   OPERATIONS RESEARCH: MATHEMATICS AND MODELS
Edited by S. I. Gass (Duluth, Minnesota, August 1979)
VOLUME 26   THE MATHEMATICS OF NETWORKS
Edited by S. A. Burr (Pittsburgh, Pennsylvania, August 1981)
VOLUME 27   COMPUTED TOMOGRAPHY
Edited by L. A. Shepp (Cincinnati, Ohio, January 1982)
VOLUME 28   STATISTICAL DATA ANALYSIS
Edited by R. Gnanadesikan (Toronto, Ontario, August 1982)
VOLUME 29   APPLIED CRYPTOLOGY, CRYPTOGRAPHIC PROTOCOLS, AND COMPUTER SECURITY MODELS
By R. A. DeMillo, G. I. Davida, D. P. Dobkin, M. A. Harrison, and R. J. Lipton (San Francisco, California, January 1981)
VOLUME 30   POPULATION BIOLOGY
Edited by Simon A. Levin (Albany, New York, August 1983)
VOLUME 31   COMPUTER COMMUNICATIONS
Edited by B. Gopinath (Denver, Colorado, January 1983)

AMS SHORT COURSE LECTURE NOTES
Introductory Survey Lectures
published as a subseries of
Proceedings of Symposia in Applied Mathematics

**PROCEEDINGS OF SYMPOSIA
IN APPLIED MATHEMATICS**
Volume 32

Environmental and Natural Resource Mathematics

AMERICAN MATHEMATICAL SOCIETY
PROVIDENCE, RHODE ISLAND

LECTURE NOTES PREPARED FOR THE
AMERICAN MATHEMATICAL SOCIETY SHORT COURSE

ENVIRONMENTAL AND NATURAL RESOURCE MATHEMATICS

HELD IN EUGENE, OREGON
AUGUST 14–15, 1984

EDITED BY
ROBERT W. McKELVEY

The AMS Short Course Series is sponsored by the Society's Committee on Employment and Education Policy (CEEP). The series is under the direction of the Short Course Advisory Subcommittee of CEEP.

Library of Congress Cataloging in Publication Data
Main entry under title:
Environmental and natural resource mathematics.
　　(Proceedings of symposia in applied mathematics; v. 32. AMS short course lecture notes)
　　　　Includes bibliographies.
　　　　1. Environmental policy—Mathematics—Congresses. 2. Environmental protection—Mathematics—Congresses. 3. Natural resources—Management—Mathematics—Congresses.
I. McKelvey, Robert W., 1929–   . II. American Mathematical Society. III. Proceedings of symposia in applied mathematics; v. 32. IV. Proceedings of symposia in applied mathematics. AMS short course lecture notes.
HC79.E575　　1985　　　　　　　　　　33.7′01′51　　　　　　　　85-3917
ISBN 0-8218-0087-6 (alk. paper)

　　COPYING AND REPRINTING. Individual readers of this publication, and nonprofit libraries acting for them, are permitted to make fair use of the material, such as to copy an article for use in teaching or research. Permission is granted to quote brief passages from this publication in reviews, provided the customary acknowledgement of the source is given.
　　Republication, systematic copying, or multiple production of any material in this publication (including abstracts) is permitted only under license from the American Mathematical Society. Requests for such permission should be addressed to the Executive Director, American Mathematical Society, P.O. Box 6248, Providence, Rhode Island 02940.
　　The appearance of the code on the first page of an article in this book indicates the copyright owner's consent for copying beyond that permitted by Sections 107 or 108 of the U.S. Copyright Law, provided that the fee of $1.00 plus $.25 per page for each copy be paid directly to the Copyright Clearance Center, Inc., 21 Congress Street, Salem, Massachusetts 01970. This consent does not extend to other kinds of copying, such as copying for general distribution, for advertising or promotional purposes, for creating new collective works, or for resale.

1980 Mathematics Subject Classification. Primary 49-06, 90-06, 92-06.
Copyright © 1985 by the American Mathematical Society.
Printed in the United States of America.
This volume was printed directly from copy prepared by the authors.
The paper used in this book is acid-free and falls within the guidelines
established to ensure permanence and durability.

CONTENTS

Preface ix

Applications of mathematics in insect pest management
    RICHARD E. PLANT 1

Economic incentives for pollution control
    MAUREEN L. CROPPER 19

Depletion and discounting: a classical issue in the economics of exhaustible resources
    GEOFFREY HEAL 33

Capital theoretic aspects of renewable resource management
    COLIN W. CLARK 45

Applying abstract control theory to concrete models
    FRANK H. CLARKE 55

International trade in resources: a general equilibrium analysis
    GRACIELA CHICHILNISKY 75

The role of mathematicians in natural resource modeling
    PANEL DISCUSSION 127

PREFACE

This volume contains lecture notes, mildly revised from those originally prepared for the American Mathematical Society's 1984 Summer Short Course, held in Eugene, Oregon on August 14-15. It also contains an expanded version of a panel discussion at the Short Course, on the role played by mathematicians in natural resource modeling.

The term "natural resources" is to be interpreted broadly, encompassing air and water resources, land and soil, minerals and oil, energy resources, and such biological resources as fisheries, agricultural crops, forests, and wildlife. The objective of the Short Course, and of this volume, is to demonstrate that, despite the great diversity of kinds of natural resources, there has developed a coherent theory concerning the efficient and conservative management of resources, and that this theory has a substantial mathematical component.

The first lecture, by Richard Plant, is a case study, introducing us to the field of agricultural insect pest management. Several management issues are investigated, relating to the economically optimal timing of the application of chemical pesticides, to the maintenance within the insect population of susceptibility to the pesticides, and to the use of a biological control technique, namely the release of sterile males into an indigenous pest population. These issues are explored through simple, analytically tractable, dynamic population models. These, despite their simplifying assumptions, provide insight into the effects of both environmental and biological factors, some of which are subject to management control. Stochastic variability is demonstrated to influence profoundly the behavior of the modeled system, and hence the appropriate management strategies. While conveying something of the flavor of a particular applied resource management field, Plant's lecture also demonstrates the spirit in which analytical resource modeling exercises are conducted generally.

In Maureen Cropper's lecture we are introduced, in the context of a water quality problem, to one of the central themes of natural resource theory: the development of strategies to cope with environmental or resource "externalities." Here, certain users of a lake or stream pollute its waters, degrading them for use by others. Technical means exist for abating or mitigating the pollution, at costs below the value to society of the improvements thus attained. The problem, then, is to work out cooperative arrangements among the parties to achieve optimal abatement, either through private bargaining or through government arranged incentives.

What makes the problem difficult is the high cost or even unavailability of information to achieve optimal abatement: in this case, information about emission abatement costs, damages from ambient pollution, and the physical relation between emission and ambient levels. Furthermore, polluting firms have an incentive to misrepresent their abatement costs and injured parties to exaggerate damages. The problem of inducing agents to truthfully reveal information is a general one in economic theory. Professor Cropper describes, through a simple model applied to the water pollution example, mechanisms that have been proposed for achieving this aim.

A classical problem associated with the management of an exhaustible resource, such as an underground pool of oil, is the determination of the appropriate rate of depletion: that is, of the time profile for the exhaustion of the resource stock. The aim is to do this in a way that is at the same time efficient and equitable: i.e., to provide society with in some sense the greatest possible value from the asset, and at the same time to achieve a fair distribution of the benefits among the generations.

These considerations are reflected in the choice of an intergenerational utility function. Very frequently the utility measure incorporates the principle of the discounting of future returns. But discounting is highly controversial, and indeed often is held to be unacceptable ethically.

Geoffrey Heal, in his lecture, demonstrates some of the paradoxical results of intertemporal utility theory. He shows how, on the basis of reasonable, and seemingly innocuous, postulates concerning intergenerational social preference orderings, one is led automatically to the discounting of future returns. He also examines the implications of uncertainty about future technological progress, and its relation to the problem of resource conservation.

For a self-renewing resource, such as a reproducing biological population, the task of resource conservation is akin to that of determining the optimal rates of investment in and consumption of an ordinary capital asset. Colin Clark elaborates on this capital-theoretic perspective on renewable resource management, through a series of optimal control models of the harvest of a marine fish stock. These explore, among others, the limitations of myopic decision rules, the complications that arise in multispecies fisheries, the effect of irreversible capital investment in fishing vessels and, again, the profound implications of uncertainty and incomplete information. Indeed much current research is concentrated on the stochastic fishery, where fish stocks, market prices and costs fluctuate erratically, and stock levels at any given time are known only indirectly and very imprecisely.

As several of the lectures demonstrate, the methods and ideas of optimal control theory play a central role in the theory of management of natural resources. In his lecture, Frank Clarke undertakes the ambitious task of surveying for the applied resource modeler some of the main results, techniques, and principles involved in analysing specific problems of optimal control. The illustrative examples are drawn largely from resource management and are intended to display a gamut of techniques. Clarke's announced intention is to make several "megapoints":

"(a) It is necessary to have a working knowledge of a variety of abstract machinery, both of its scope and its limitations.
(b) There are pitfalls associated with incomplete analysis of problems.
(c) Non-trivial optimal control problems are hard. (Yes, this may be tautological.)"

The final lecture in this volume is that of Graciela Chichilnisky. Her focus is on the macroeconomic effects of international trade in extractive resources, for example the current world trade in oil. The discussion is based on a simple model of "North-South" trade in resources, with an inflow of foreign capital to the producing country. (The recent transfer of 100 billion US dollars to Mexico is a relevant example.) The model is a condensed general equilibrium model, restricted to contain only five markets and four economic agents or groups. An unusual feature is that this model is mathematically tractable, and not merely accessible through computer simulation. It can be used not only to ascertain existence and uniqueness of the equilibrium, but to make qualitative judgments concerning a wide variety of policy issues, the results appearing in the form of theorems involving the shapes and global properties of solution manifolds, as the underlying parameters vary. The results obtained are somewhat surprising, confirming the view that the behavior of several markets, all interacting with each other, is not readily understood by the common wisdom usual to a single market. Cross market effects, the very topic of general equilibrium theory, yield the most surprising results and those which are perhaps the most useful for policy.

The Short Course concluded with a panel discussion, featuring lecturers and with participation by their audience. The discussion was taped, transcribed and edited for inclusion in this volume. The discussion explored the fascinating role that mathematicians and mathematically trained scientists have played throughout the development of the discipline of natural resource modeling, and in economic theory generally. The discussion then turned to consideration of ways in which concepts and techniques of modeling might best be incorporated into graduate and undergraduate mathematics education.

> Robert McKelvey
> Department of Mathematical Sciences
> University of Montana

APPLICATIONS OF MATHEMATICS IN INSECT PEST MANAGEMENT

Richard E. Plant[1]

ABSTRACT. The theory of insect pest management touches the theory of resource management in two ways. The first is the most direct. Insects are a pest of biological resources, and therefore a part of the management of these resources is the management of insects. The second way is more indirect. In controlling an insect population, one or more resources must frequently be consumed or altered. For example, pesticide residues may contaminate groundwater. This lecture is divided into three parts. The first two each focus on one of these two theoretical problems. The first part deals with the problem of controlling an insect population in an agricultural crop during a single season. The second part focuses on the problem of resource consumption in pest control. The third part deals with a promising nonchemical means of insect control, the sterile insect release method.

1. INTRODUCTION. Insects are a major pest of virtually every land based biological resource. Among modelers of renewable resource problems, perhaps the best known insect pest is the spruce budworm, which periodically defoliates large areas of softwood forests in Canada. The best known agricultural pest is probably the boll weevil, which invaded Texas from Mexico in 1892 and swept through the cotton belt, causing a convulsive change in the economy of the South. Agricultural commodities are not usually considered renewable resources since they do not renew themselves through an escapement, but the development of the theory of insect pest management in forests has followed essentially the same lines as that for crops. Since I am more familiar with the latter theory, I shall focus on that in this lecture.

Pest control practices may be divided into three categories. These are chemical, such as the application of pesticides; biological, such as the release of predators or parasites; and cultural, such as the physical destruction of the pests or the rotation of crops. Prior to World War II there were few effective chemical pesticides available, and farmers had to rely on all three methods, together with a certain amount of luck, for insect pest control.

---

[1] 1980 Mathematics Subject Classification. 92.02
[1] Supported by NSF Grant MSC 81-21413.

During the war, the compound DDT was discovered by Ciba-Geigy chemists and developed in the U.S. as a delousing agent for troops. Shortly after the war, the potential of DDT and related compounds for insect pest control was recognized, and these compounds were quickly put to use as all purpose insecticides. In the present era of concern over the deleterious effects of pesticides, it is difficult to recall the enthusiasm that greeted the introduction of these compounds. Many entomologists sincerely questioned whether their profession could remain viable, since it appeared that all insect pests would be quickly eradicated (Metcalf, 1980).

Farmers rapidly abandoned biological and cultural practices in favor of complete reliance on chemical pesticides. Problems, however, soon began to appear (Metcalf, 1980). Many pest species developed with amazing speed resistance to the pesticides. In addition, the pesticides frequently killed off predators of the pests as well as the pests themselves. This had two effects. The first was that after a pesticide application, pest populations would often rapidly rebound to levels higher than those before the application. The second effect was that species that had previously been unimportant, because they had been kept in check by natural enemies, suddenly emerged as serious pests. Finally, the consequences of the accumulation of pesticide residues in the environment began to appear. Thus, in the 1950's entomologists began to move back to the consideration of pest control methods that did not rely so strongly on chemical pesticides.

A major contribution was the publication in 1959 of an article by Stern et al. (1959) that synthesized many of the important concepts of what is now called integrated pest management, or IPM. The fundamental idea behind IPM is that each crop in a given area should be considered as a managed ecosystem. Pest management plans are developed in which a combination of chemical, biological, and cultural practices are used in a way appropriate for the particular crop to provide a reasonable level of control. Entomologists often use mathematical models to aid in the development of these plans (e.g., Huffaker (1980).

Serious mathematical investigation of problems arising from insect pest management began during the 1960's. One of the first workers to study these problems was Watt (1963). He used a dynamic programming approach; a similar approach was taken by Shoemaker (1973). Other early investigators studied the problem using optimal control theory (Goh et al. 1974), birth-death processes (Becker, 1970, Getz, 1975), and decision theory (Mann, 1971). Although there is an overwhelming statistical literature involving insect pest management problems, these problems have not attracted as much attention from mathematicians. The literature has, however, grown to the point where several reviews have recently appeared (Shoemaker, 1981, Feldman and Curry, 1982, Getz and

Gutierrez, 1982, Plant and Mangel, 1984). The reader should consult these for a more complete guide to this literature. Many of the theoretical concepts of insect pest management resemble, at least in a general way, those of epidemiology. This field has a much more well developed literature, reviewed, for example in (Wickwire, 1977).

This lecture will attempt to describe the development of the mathematical theory of insect pest management within the IPM context. The theory will be presented in as general a manner as possible. We will, however, focus on one particular crop, cotton, in one particular region, the San Joaquin Valley of California. This will add definiteness and will, I hope, provide a little more of the flavor of the field.

Section 2 deals with the problem of managing a crop over the course of a single season. The main topic of this section is the economic threshold, which is a fundamental part of the theory of integrated pest management. Section 3 considers the long term management problem. In particular, we examine strategies to deal with the development of resistance to a pesticide in the pest population. Section 4 discusses the sterile insect release method, which is a good example of a modern nonchemical approach to pest management.

2. THE ECONOMIC THRESHOLD. In this section we consider the control problem faced by the grower of an agricultural crop. The grower manages the crop over the course of a season, but all of his reward comes from the yield that he harvests at the season's end. Farming is well known as a business fraught with uncertainty and risk, so it would be reasonable to phrase the grower's problem in the context of stochastic optimal control. Much of the theory associated with this problem has, however, been developed using a deterministic model. We therefore first formulate the problem in a deterministic way and then add stochastic elements to it. A major focus of integrated pest management has been to develop a simple rule that growers can use to determine when to apply a pesticide. We will show how this rule, which is called the <u>economic threshold</u>, may be derived and implemented.

Throughout the discussion, we shall carry along the example of the San Joaquin Valley cotton agroecosystem, and use it for purposes of illustration. In describing the example, we will not in general cite specific sources. These sources include Toscano et al., 1979, Huffaker, 1980, Flint and van den Bosch, 1981, Anon., 1984, and personal discussions with T. Dennehy, J. Granett, T. Leigh, and L. T. Wilson.

The major insect pest species in the San Joaquin Valley cotton agroecosystem are lygus bugs and spider mites. Lygus bugs are a little bigger than a ladybug, brown, and flat. Spider mites are not really insects at all, but rather arachnids. They are so tiny that they are hard to see with the naked eye. The major predators of lygus bugs are big eyed bugs and damsel bugs. The major predators of mites are insects called thrips.

Before the advent of integrated pest management, many pesticide applications were made without much concern for the actual state of the pest infestation. This often resulted the application of a considerably larger quantity of pesticide, with all the associated consequences, than actually needed. For example, in the San Joaquin Valley, the unnecessary application of insecticide to kill lygus bugs also greatly reduced the thrip population, resulting in heavy outbreaks of spider mites.

The economic threshold concept, which is generally attributed to Stern et al. (1959), is intended to give the grower a rational means of determining when the pesticide should be applied. Stern et al. define the economic threshold to be the level of pest population at which application of pesticide becomes economically justified. The first effort to mathematically define the economic threshold was made by Headley (1972), and later generalized by Hall and Norgaard (1973). We begin our discussion with the Headley-Hall-Norgaard model. In the San Joaquin Valley during many seasons lygus bugs are not a major concern, so in discussing applications we shall focus our attention on spider mites.

The model consists of the following five elements: a pest population growth function $x(t)$; a pesticide induced mortality rate $K(u)$, where u is the quantity of pesticide applied; a crop damage function D; and a pesticide cost function $C(u)$. Hall and Norgaard assume linear pest population dynamics with a fixed immigration rate, so that their population equation is

$$x(t) = \begin{cases} x_0 e^{rt} + q: & 0 \leq t \leq \tau \\ (x_0 e^{r\tau}[1-k])e^{r(t-\tau)} + q: & \tau < t \leq T \end{cases} \qquad (2.1)$$

where the season begins at $t = 0$, the single pesticide application is made at $t = \tau$, and the crop is harvested at $t = T$. The assumption that the rate of damage to the crop is proportional to the pest population yields

$$D(\tau, u) = \int_0^T b x(t) \, dt. \qquad (2.2)$$

Let N denote the yield if no damage to the crop had occurred. Assume that the cost of pesticide application is given by

$$C(u) = \varphi + \alpha u \qquad (2.3)$$

where φ is the fixed cost and α is the unit cost associated with an application. Let β denote the unit price of the crop. Then the profit π to the grower is given by

$$\pi(\tau,u) = \beta[N-D(\tau,u)] - C(u). \qquad (2.4)$$

The problem is to choose τ and u to maximize $\pi(\tau,u)$. Assuming an interior maximum, this is simply a calculus problem. Hall and Norgaard (1973, 1974) and Borosh and Talpaz (1974) give a discussion of the conditions on $K(u)$ necessary and sufficient for the existence of an interior maximum.

Having formulated the initial model, let us consider some of the assumptions involved in the formulation. Equation (2.1) contains the assumption that there are no density dependent effects, which is reasonable since the grower would not let the population grow large enough for such effects to become important. The equation does not explicitly include natural controls such as thrips as a control of mites. The effect of such controls may, however, be included in the intrinsic growth rate r. The equation also does not explicitly include the effects of temperature; these effects are important in insect populations. Agriculturists measure time not in chronological, but rather in "physiological" units (Wilson and Barnett, 1983). Physiological time is the integral over chronological time of the variable $T'(t)$, where

$$T'(t) = \begin{cases} T(t) & \text{if } T_1 < T(t) < T_u \\ 0 & \text{otherwise.} \end{cases} \qquad (2.5)$$

The thresholds T_1 and T_u are the temperatures at which the growth process ceases. We shall include the effects of temperature by assuming that time is measured in physiological units.

Another objection to the Headley-Hall-Norgaard model is that it only provides for a single application of pesticide per season. Talpaz and Borosh (1974) have extended the model to cover multiple applications by assuming that the applications are evenly spaced.

In the case of spider mites in San Joaquin Valley cotton, one application of pesticide is frequently all that is needed, so one would think that the Headley-Hall-Norgaard model would serve as a useful paradigm for an IPM program. In fact it does not, nor does it serve this role in any other significant agroecosystem. Instead, entomologists generally recommend, and growers generally practice, the following program. At each time at which he measures the pest population, the grower compares the cost of applying the pesticide at the maximum level allowed by law (pesticide application rates are regulated under the Federal Insecticide, Fungicide, and Rodenticide Act, or FIFRA) with the projected cost of the crop damage if no pesticide is applied. If the latter cost is greater than the former, then the grower applies the maximum allowable amount of pesticide.

Hall and Moffit (1984) discuss this difference between the control program recommended by economists and that recommended by entomologists. They point out that the latter program is economically irrational in the sense that it does not maximize the grower's profit. This apparent irrationality may, however, be understood by noting that the Headley-Hall-Norgaard model does not take into account stochastic effects. Plant (1984) has recently discussed the effect of uncertainty on the model. We sketch this discussion here.

The model is reformulated in terms of stochastic optimal control theory. In this formulation, the grower at regular intervals measures the level of infestation in the crop, and, if these measurements indicate that a response is appropriate, applies a pesticide. Since stochastic control problems are most easily formulated in discrete time, we let $x_n = x(t_n)$ where $t_n = nh$, h being a fixed time interval, and go to a discrete time setting. In the absence of pesticide application, the linear dynamics of equation (2.1) imply that we may write $x_{n+1} = \rho x_n$ where $\rho = e^{rh}$. We assume that the fraction K of insects killed by a pesticide application is a stochastic function of the level u_n of pesticide applied in period n. We may therefore write

$$x_{n+1} = \rho(1-K(u;\xi_n))x_n + q, \qquad (2.6)$$

where $K(u;\xi)$ is a function of the control level u and the random variable ξ_n. $K(u;\xi)$ has the property that $K(0;\xi) = 0$ and $\lim_{u \to \infty} K(u;\xi) = 1$ for all ξ.

The profit function of equation (2.4) is then replaced by the function

$$\pi(x_0, \Gamma) = E\{N - \sum_{n=0}^{N} (bx_n + C(u_n))\} \qquad (2.7)$$

where x_n is defined by equation (2.6), the random variable ξ_n are independent and identically distributed, and the cost of pesticide application at stage n is given by

$$D(u) = \begin{cases} 0: & u = 0 \\ \varphi + \alpha u: & u > 0. \end{cases} \qquad (2.8)$$

In equation (2.7), Γ represents a particular sequence of pesticide levels $\{u_0, u_1, \ldots, u_{N-1}\}$ and is called a policy. Our problem is then to search for the policy Γ^* that maximizes (or, more precisely, causes to attain its supremum) the expected profit function $\pi(x_0, \Gamma)$. This problem may be solved by stochastic dynamic programming (e.g. Bertsekas 1976).

The method of stochastic dynamic programming rests on the "principle of optimality" which says that if one has computed an optimal subsequence $\{u_m^*(x), \ldots, u_{N-1}^*(x)\}$ then one may compute $u_{m-1}^*(x)$ as that function which maximizes $\pi_{m-1}(x)$ where

$$\pi_{m-1}(x) = E\{-bx - C(u_{m-1}) + \pi_m^*(\rho[1-K(u_{m-1};\xi_{m-1})]x)\} \tag{2.9}$$

and $\pi_m^*(x)$ is the profit obtained by applying the policy $\{u_m^*(x),\ldots,u_{N-1}^*(x)\}$ and summing the right hand side of equation (2.7) from m to N. One may obtain the optimal policy inductively by noting that $\pi_{N+1}^*(x) = 0$.

For purposes of exposition we will only consider the problem at stage N-1. We assume that the ξ_n are normally distributed with mean zero and variance σ^2 and that the function $K(u;\xi)$ is given by $K(u;\xi) = K_d(ue^\xi)$. We have $\pi_N(x) = bx$, and

$$\begin{aligned}\pi_{N-1}(x) &= \max_u E\{-bx - C(u) + \pi(\rho(1-K(u;\xi))x)\} \\ &= \max[N - b(1+\rho)x - (\varphi+\alpha u) + b\rho\overline{K}(u)x] \end{aligned} \tag{2.10}$$

where $\overline{K}(u) = E\{K(u;\xi)\}$. For $\varphi > 0$ the value u_{N-1}^* that maximizes the right hand side of equation (2.10) will be zero for sufficiently small P. There is a value x_c such that $u_1^* = 0$ for $x < x_c$ and $u_1^* = u_c > 0$ for $x = x_c$. The quantity x_c is the economic threshold for period N-1. Moreover, as the variance σ^2 increases, the value of u_c increases provided α is small relative to φ. The economic interpretation of this is that the grower, if he is a profit maximizer, will tend to apply more pesticide in the presence of uncertainty than he would otherwise. Since there is a maximum permissible amount u_{max} set by federal regulations, if $u_c > u_{max}$ then the grower will apply an amount u_{max}. At this point the economist's policy and the entomologist's policy become equivalent.

Our conclusion is that, given the present level of uncertainty associated with the various parameters of the insect pest control problem, the grower may not be acting irrationally at all when he applies pesticide at the maximal rate. He may simply be taking into account the effects of uncertainty.

3. PESTICIDE RESISTANCE. When the grower applies a pesticide to control a population of insects in his crop, it may happen that a tiny fraction of the insects are resistant to the effects of the pesticide. If this is the case, then these individuals will have a higher survival rate and therefore a greater chance of leaving progeny for the next generation. If the resistance is genetically conferred, then each succeeding application of the pesticide will select for resistant individuals, and the fraction, or frequency, of resistant individuals will increase. Eventually the pesticide may lose its ability to control the pest population. This transition to ineffectiveness may take place in just a few years (Georghiou, 1980). There is evidence (Dennehy and Grannet, 1984) that the spider mite population in California is developing in this way resistance to dicofol, one of the primary mite control agents.

Since resistance is a genetic phenomenon, it should be amenable to study by the methods of population genetics. In many cases resistance is governed primarily by the simple single locus, two allele framework that is the subject of most population genetics theory (Brown, 1971). In this section we discuss this theory and its application to pest management.

In the single locus, two allele model, resistance is conferred by a single allele, denoted R; the other allele, which confers susceptibility, is denoted S. Each individual carries a pair of these alleles, either RR, RS, or SS. Individuals with the pair RR (homozygous resistants) are resistant, those with the pair SS (homozygous susceptibles) are susceptible, and, depending on the level of dominance of the R allele, those with RS (heterozygotes) are somewhere in between.

The classical Hardy-Weinberg formula (e.g., Ewens, 1968) describes the evolution of the frequencies of the two alleles. It is based on several restrictive assumptions, including discrete, nonoverlapping generations; monoecious reproduction, and random mating. Nevertheless, this law often works well for systems that violate one or more of these assumptions, and it is commonly employed as a model for these systems. The law, as we will use it, may be stated as follows. Assume that in the nth generation the RR, RS, and SS genotypes have survivorships A_n, B_n, and C_n, respectively. Let p_n denote the frequency of the R allele, and q_n ($= 1 - p_n$) denote the frequency of the S allele. Then according to the Hardy-Weinberg formula p_n evolves according to the equation

$$p_{n+1} = \frac{A_n p_n^2 + B_n p_n q_n}{A_n p_n^2 + 2B_n p_n q_n + C_n q_n^2}. \tag{3.1}$$

For simplicity, let the function of p_n on the right hand side of equation (3.1) be denoted $F_n(p_n)$. We assume that if a pesticide is applied in generation n, then $A_n \gg C_n$, and B_n is between A_n and C_n.

The application of population genetics theory to pesticide resistance has been considered by a number of workers (e.g., Hueth and Regev 1974, Taylor and Headley 1975, and Shoemaker 1982). In this lecture we shall focus on two particular models, one due to Comins (1977a,b) and one to Georghiou and Taylor (1977a,b). Georghiou and Taylor (1977a) study an approximation to equations (3.1) in which they assume that no members of the SS genotype survive the spraying, so that $C_n \equiv 0$. They construct a simulation model in which the population of each genotype obeys a growth equation of the form

$$x_{i,n+1} = x_{i,n} \exp(r[K-x_{i,n}]/K), \tag{3.2}$$

with i = RR, RS, and SS. The genotypes interact through equation (3.1), with

$A_n = W_{RR} x_{RR,n}$, and similarly for B_n and C_n. Using their simulation model, Georghiou and Taylor (1977a,b) study the effects of controllable factors such as pesticide dose, and uncontrollable factors such as degree of dominance of the R allele, on the rate of evolution of resistance. They conclude that in the presence of migration, the buildup of resistance may be delayed or stopped by the maintenance of refugia, i.e., of locations where a portion of the population is left untreated. They also find that the growth resistance is strongly affected by the degree of dominance of the resistance conferring allele.

Comins (1977a) also studies the effect of migration on the development of resistance. He considers the case in which only the larvae, which do not reproduce and are relatively immobile, are treated with insecticide. This is often true with lepidopterous pests (i.e., caterpillars) but it is unfortunately not true of spider mites, which are pests in the adult stage. Spider mites do not have a larval stage, but the juvenile stage is relatively immobile. We shall review Comins' work with the idea that, like the Hardy-Weinberg law, it may be applicable to systems beyond the range of its assumptions. If we accept that only larvae are treated, then both migration and reproduction occur at times distinct from pesticide treatment. Comins considers a system shown schematically in Figure 3.1.

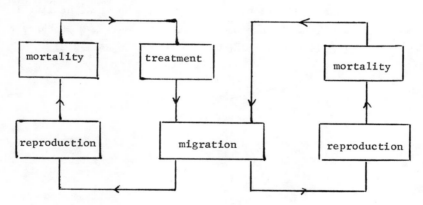

Fig. 3.1

The system consists of a treated "pool" with population x_n at generation n, and an untreated "pool" with population y. For simplicity we will consider only the case in which the untreated pool is sufficiently large that its population may be regarded as fixed. Assume that in the absence of migration the total population obeys a dynamical relationship of the form

$$x_{n+1} = \lambda x_n^{1-b} . \qquad (3.3)$$

This system has fixed points at $x = 0$ and at $x = \lambda^{1/b}$. For $\lambda > 0$ and $0 < b < 2$ the equilibrium at $x = 0$ is unstable and that at $x = \lambda^{1/b}$ is stable. Assume

that the effect of immigration is to add a fixed number M of individuals per
generation to the treated pool. Let the (fixed) frequency of the resistance
allele in the untreated population be denoted w. Assume further that the fit-
nesses terms of equation (3.1) are independent of n, and that $B = hA + (1-h)C$
where $0 \leq h \leq 1$. The equations describing the dynamics of the population x_n
and the frequency p_n of the resistance allele may then be written

$$x_{n+1} = (1-M)K(p_n)x_n^{1/b} + M$$
$$x_{n+1}p_{n+1} = (1-M)L(p_n)x_n^{1-b} + Mw, \quad (3.4)$$

where $K(p) = Ap^2 + 2Bp(1-p) + C(1-p)^2$, and $L(p) = Ap^2 + Bp(1-p)$. After a
little algebra, one can show that an equilibrium solution $(\bar{x},\bar{p})$ of this system
must satisfy

$$\bar{p}(1-\bar{p})[h+(1-2h)\bar{p}] = s(\bar{p}-w)/\bar{x}^{1-b} \quad (3.5)$$

where

$$s = M/[(1-M)(A-C)]. \quad (3.6)$$

The case b = 1 is particularly simple. In this case the equilibrium solutions
are given by the intersections of the cubic on the left side of equation (3.6)
with the straight line on the right. The slope of the straight line is s and
its intercept is w. If we allow the migration parameter M to vary then we
obtain the locus of equilibrium solutions $\bar{p}$ shown in Fig. 3.2.

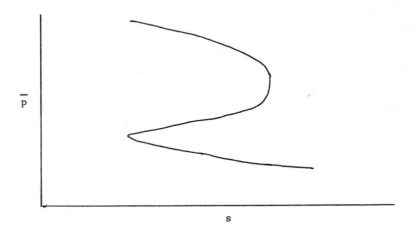

Fig. 3.2

The values M_1 and M_2 are branch points; the stability of the equilibrium
branches is shown in the figure. For sufficiently small M there is only a
large valued equilibrium; for intermediate M there are two stable equilibria
separated by an unstable one, and for large M there is a unique, small valued

equilibrium. The conclusion is that migration from a sufficiently large pool of untreated individuals may have a significant effect in delaying the buildup of resistance.

In a later paper, Comins (1977b) examines the case in which there is no migration to try to determine optimal pesticide application strategies. He considers the case in which $p_n \ll 1$, since if $p_n = 0(1)$ the pesticide would not be used. Expanding the function $F_n(p_n)$ of equation (3.1) about $p_n = 0$ yields

$$F_n(p_n) = \frac{B_n}{C_n} p_n + o(p_n). \tag{3.7}$$

If we assume again that each generation is treated with the same dose of pesticide, then equation (3.7) implies that the frequency of the resistance gene grows at a geometric rate as long as p_n is small. If the R allele is dominant then $B \gg C$ and, as observed in the simulations of Georghiou and Taylor, the rate of increase is high. A high degree of dominance by the R gene therefore affects the rate of buildup of resistance both by conferring resistance on the RS as well as the RR genotype and by speeding the rate of growth of the R allele's frequency. On the other hand, the degree of resistance of the homozygous population has, to first order, no effect on the growth of resistance.

To get a simple form for the equations, Comins alters the model slightly from that of his previous paper. He assumes that the survivorship of the SS genotype in the ith generation is s, and the survivorship of the RS genotype is $s_i^{1-\beta}$. Taking logarithms of both sides of equation (3.7) implies that in the case of a pest with m generations per year, the change in $\ln p$, the logarithm of the resistance gene frequency, may be written

$$\Delta \ln p = \beta |\ln S|, \tag{3.8}$$

where $S = s_1 s_2 \ldots s_m$. S is called the "survival product", and if it is assumed to be the same for each year, then the time for resistance to develop will be proportional to $1/|\ln S|$.

To obtain an optimal strategy for pesticide application, Comins assumes that the cost a_i of application of the pesticide in the ith generation may be written

$$a_i = A_0 |\ln s_i| \tag{3.9}$$

and that the damage d_i done to the crop by this generation has the form

$$d_i = s_i D(d_{i-1}/D)^{1-b}, \tag{3.10}$$

where D is the amount of damage done by a pest population at its equilibrium state $\lambda^{1/b}$. The total cost per year of the pesticide strategy is therefore

$$C = \Sigma d_i + f(\ln S) + A_0 |\ln S|. \qquad (3.11)$$

The problem is therefore to minimize this cost. Comins presents some solutions to this problem obtained using Lagrange multipliers. He finds that, qualitatively speaking, the best strategy is to keep the damage done by each generation roughly the same.

Other workers (e.g., Gutierrez et al., 1979, Shoemaker, 1982, Plant et al. 1984) have also studied the long term management problem. This is an area, however, in which much work remains to be done. One of the most challenging problems is that, due to the very low initial value of p_n and the geometric growth of resistance in its early stages (equation [3.7]), the increase in resistance in a population is often not detected until it is almost too late to do anything about it.

4. THE STERILE INSECT RELEASE METHOD. The use of chemical pesticides to control insect pests consumes resources at several ecological levels. Entomologists have been actively seeking alternatives to the so-called "broad spectrum" insecticides. These alternative means of pest management are often best utilized as a complement to chemical pesticides, and indeed the fundamental tenet of IPM is to intelligently use a combination of all available pest management techniques.

One technique that has received considerable publicity of late is the sterile insect release method (SIRM). In this method, large numbers of artificially sterilized adults are released into the pest population. These sterilized individuals mate with the native members of the population and thereby reduce the population's total offspring. Eventually, the population may be controlled or even eradicated by this method.

This method is only appropriate for insect species in which the adult stage does not do a significant amount of damage. If the members of the species can be easily separated by sex, and if the adult male does not cause damage, then sterilized adult males can be released. There are a variety of other criteria that must be satisfied for the SIRM to work. Unfortunately, the two major cotton pests in the San Joaquin Valley, the spider mite and the lygus bug, fail to meet these criteria.

There are, however, two well known pests of cotton that are extending their range in the direction of the San Joaquin Valley, the pink bollworm and the boll weevil. One or the other or both of these cause major damage to cotton in virtually every other cotton growing region in the United States. The SIRM has been employed in an effort to control each of these species.

The SIRM was first proposed by Knipling (1955), who employed a simple mathematical model to indicate how effective it might be. His model is as follows. Suppose that in the absence of treatment, the pest population obeys the linear dynamical relation

$$x_{n+1} = \rho x_n, \qquad (4.1)$$

where, as usual, x_n is the population in generation n, and ρ is the intrinsic growth rate. Suppose further that s sterile individuals are released in each generation. If these individuals are equally competitive with the fertile members of the population, then we may model the dynamics of the population with the equation

$$x_{n+1} = \frac{\rho x_n^2}{x_n + s}. \qquad (4.2)$$

This equation has an unstable equilibrium at $x = s/(r-1)$. Therefore, if $x_0 < s/(r-1)$ the x_n will approach zero. Moreover, the per generation decay rate will be $\rho x_n/(x_n+s)$, so that the population will decline at a greater than geometric rate. Because of this, eradication can, in principle, be achieved in just a few generations.

Ito and his coworkers (Ito, 1977; Ito and Kawamoto, 1979; Ito and Koyama, 1982) use the SIRM model to simulate a campaign conducted to eradicate the melon fly from the island of Kume in Japan. Since the melon fly had been present on this island for many years, there was reason to believe that density dependent effects were important in regulating its population. Ito uses the equation

$$x_{n+1} = \left[\frac{\rho K x_n}{K + (1-r)x_n}\right] \qquad (4.3)$$

to model the density dependent growth dynamics; immigration may be considered zero for an island population. For simplicity let the right hand side of (4.3) be denoted $G(x_n)$. Ito develops a more complicated, and presumably more realistic, model in place of equation (4.2) for the effect of sterile individuals on reproduction. Moreover, he assumes that the sterile population affects only the growth term of the model, and not the density dependence term. In other words, he writes

$$x_{n+1} = G(x_n)H(x_n,s) \qquad (4.4)$$

where $H(x,s)$ represents the effects of the sterile insects.

Ito assumes that female flies mate between zero and six times in their lifetimes, with the number of matings being a poisson distributed variable. He assumes that 20 percent of the population do not mate at all; this assumption determines the poisson parameter λ. Let

$$P_j = \frac{\lambda^j}{j!} e^\lambda, \quad j=0,\ldots,5, \tag{4.5}$$

and let $P_6 = 1 - \Sigma P_j$, the sum being from 0 to 5. Our representation of Ito's model will be slightly different from his in that he absorbs into another parameter the effect of flies who do not mate at all. Ito, following Berryman (1967) assumes that the event of mating with a fertile or a sterile individual is binomially distributed. The effect of multiple mating in the model is as follows. Females who mate with only sterile males produce viable eggs with probability zero, those who mate with only fertile males produce viable eggs with probability one, and those who mate with a combination of sterile and fertile males produce viable eggs with probability 1/2, independent of the combination. The mating function $H(x,s)$ may then be written

$$H(x,s) = \sum_{j=0}^{6} P_j [\varphi^j + \tfrac{1}{2}\{1-(\varphi^j + [1-\varphi]^j\}], \tag{4.6}$$

where $\varphi = s/x_n$. Note that if all females were assumed to mate exactly once, then equation (4.6) would reduce to

$$H(x,s) = \frac{x}{x+s}, \tag{4.7}$$

which is the Knipling model. Ito finds that the effect of multiple mating on the function $H(x,s)$ is quite small. Using the model, Ito and his coworkers are able to determine release rates for sterile insects on Kume island. The eradication of the melon fly from this island was announced in 1977.

Prout (1978) studies the effect of density dependence and migration on the SIRM. His model of the untreated population without migration is the same as Ito's, given by equation (4.3). Unlike Ito, however, Prout multiplies the term x_n in both the numerator and the denominator by $x_n/(x_n+s)$ to account for the release of sterile individuals. Migration is included through a parameter M, giving the equation the form

$$x_{n+1} = \frac{rK(x_n^2 + M(x_n+s))}{K(x_n+s) + (r-1)(x_n^2 + M(x_n+s))}. \tag{4.8}$$

Prout also studies other models for various forms of migration and sterile insect release, but we will focus on equation (4.8). The equilibria of this equation are the roots of the cubic

$$F(x;K,M,r,s) = rKMs \tag{4.9}$$

where

$$F(x;K,M,r,s) = (4-1)x^3 + (r-1)(M-K)x^2 + [Ms(r-1) + K(s-Mr)]x. \tag{4.10}$$

We briefly sketch the analysis of this equation; a more complete treatment is given in Plant and Mangel (1984).

If $M = 0$ (i.e., there is no migration), then $x = 0$ is always a root of this equation. If, in addition, $s > K(r-1)/4$, then this root is stable and is the only real root. This corresponds to eradication of the population. If $M > 0$ then $x = 0$ is not a root of equation (4.9). Therefore, the effect of migration on the model is to eliminate the possibility of complete eradication. If the inequality $s > K(r-1)/4$ still holds, then for sufficiently small M this equilibrium is stable and unique. As M increases a larger valued equilibrium appears and branches into a pair, with the largest value being stable. As M increases still further, the unstable and lower stable equilibria coalesce and disappear, leaving only the large valued equilibrium. The conclusion is that migration of fertile insects into the region of sterile release eliminates the possibility of eradication, and that a sufficiently large migration rate may destroy the effect of the sterile insect release program.

The predictions of Prout's model are borne out in actual field observations. The only successful eradication campaigns using the SIRM method have been those either conducted on islands, where there is no migration, or against local infestations of exotic pests. The implication for pest management in the San Joaquin Valley is that eradication of the pink bollworm and boll weevil using the SIRM is not possible. These methods may, however, be used to aid in keeping the population at a low level, provided their migration rates are not too large.

BIBLIOGRAPHY

Anonymous (1984). "Integrated Pest Management for Cotton", Preprint, University of California, Davis.

Becker, N. G. (1970), "Control of a pest population", Biometrics 26: 365-375.

Berryman, A. A. (1967), "Mathematical description of the sterile male principle", The Canadian Entomologist 99: 858-865.

Bertsekas, D. P. (1976), Dynamic Programming and Stochastic Control, Academic Press, New York, 397 pp.

Borosh, I. and H. Talpaz (1974), "On the timing and application of pesticides: comment", American Journal of Agricultural Economics 56: 642-643.

Brown, A. W. A. (1971), "Pest resistance to pesticides", In Pesticides in the Environment", R. White-Stevens, ed., vol. 1, pp. 457-552.

Comins, H. N. (1977a), "The development of resistance in the presence of migration", Journal of Theoretical Biology 64: 177-197.

Comins, H. N. (1977b), "The management of pesticide resistance", Journal of Theoretical Biology 65: 399-420.

Dennehy, T. J. and J. Granett (1984), "Monitoring dicofol resistant spider mites in California cotton", Preprint.

Ewens, W. J. (1968), Population Genetics, Methuen, London.

Feldman, R. M. and G. L. Curry (1982), "Operations research for agricultural pest management", Operations Research 30: 601-618.

Flint, M. L. and R. van den Bosch (1981), Introduction to Integrated Pest Management, Plenum Press, New York.

Georghiou, G. P. (1980), "Insecticide resistance and prospects for its management", Residue Reviews 76: 131-145.

Georghiou, G. P. and C. E. Taylor (1977a), "Genetic and biological influences in the evolution of insecticide resistance", Journal of Economic Entomology 70: 319-323.

Georghiou, G. P. and C. E. Taylor (1977b), "Operational influences in the evolution of insecticide resistance", Journal of Economic Entomology 70: 653-658.

Getz, W. M. (1975), "Optimal control of a birth and death process population model", Mathematical Biosciences 23: 87-111.

Getz, W. M. and A. P. Gutierrez (1982), "A perspective on systems analysis in crop production and insect pest management", Annual Reviews in Entomology 27: 447-466.

Goh, B. S. G. Leitmann, and T. L. Vincent (1974), "Optimal control of a prey-predator system", Mathematical Biosciences 19: 263-286.

Gutierrez, A. P., U. Regev, and H. Shalit (1979), "An economic optimization model of pesticide resistance: alfalfa and egyptian alfalfa weevil - an example", Environmental Entomology 8: 101-107.

Hall, D. C. and R. B. Norgaard (1973), "On the timing and application of pesticides", American Journal of Agricultural Economics 55: 198-201.

Hall, D. C. and R. B. Norgaard (1974), "On the timing and application of pesticides: reply", American Journal of Agricultural Economics 56: 644.

Hall, D. C. and L. J. Moffitt (1984), "Stochastically efficient economic thresholds for discrete choices", Preprint.

Headley, J. C. (1972), "Defining the economic threshold", In Pest Control Strategies for the Future, National Academy of Science, Washington, D.C., pp. 100-108.

Hueth, D. and U. Regev (1974), "Optimal pest management with increasing pest resistance", American Journal of Agricultural Economics 56: 543-552.

Huffaker, C. B., editor (1980), New Technology of Pest Control, Wiley - Interscience, New York, 500 pp.

Ito, Y. (1977), "A model of sterile insect release for eradication of the melon fly, Dacus cucurbitae Coquillet", Applied Entomology and Zoology 12: 303-312.

Ito, Y. and T. Kawamoto (1979), "Number of generations necessary to attain eradication of an insect pest with sterile insect release method: a model study", Researches on Population Ecology 20: 216-226.

Ito, Y. and J. Koyama (1982), "Eradication of the melon fly: role of population ecology in the successful implementation of the sterile insect release method", Protection Ecology 4: 1-28.

Knipling, E. F. (1955), "Possibilities of insect control or eradication through the use of sexually sterile males", Journal of Economic Entomology 48: 459-462.

Mann, S. H. (1971), "Mathematical models for the control of pest populations", Biometrics 27: 357-368.

Metcalf, R. L. (1980), "Changing role of insecticides in crop protection", Annual Reviews in Entomology 25: 219-256.

Plant, R. E. (1984), "Uncertainty and the dynamic economic threshold", Preprint.

Plant, R. E. and M. Mangel (1984), "Modeling and simulation in agricultural pest management", Preprint.

Plant, R. E., M. Mangel, and L. Flynn (1984), "Multiseasonal management of an agricultural pest II: the economic optimization problem", Preprint.

Plant, R. E. and L. T. Wilson (1984), "A Bayesian method for sequential sampling and forecasting in agricultural pest management", Preprint.

Prout, T. (1978), "The joint effects of the release of sterile males and immigration of fertilized females on a density regulated population", Theoretical Population Biology 13: 40-71.

Shoemaker, C. A. (1973), "Optimization of agricultural pest management II: formulation of a control model", Mathematical Biosciences 17: 357-365.

Shoemaker, C. A. (1981), "Applications of dynamic programming and other optimization methods in pest management", IEEE Transactions on Automatic Control AC-26: 1125-1132.

Shoemaker, C. A. (1982), "Optimal integrated control of univoltine pest populations with age structure", Operations Research 30: 40-61.

Stern, V. M., R. F. Smith, R. van den Bosch, and K. S. Hagen (1959), "The integrated control concept", Hilgardia 29: 81-101.

Talpaz, H. and I. Borosh (1974), "Strategy for pesticide use: frequency and applications", American Journal of Agricultural Economics 56: 769-775.

Taylor, C. R. and J. C. Headley (1975), "Insecticide resistance and the evaluation of control strategies for an insect population", The Canadian Entomologist 107: 237-242.

Toscano, N. C., V. Sevacherian, and R. A. Van Steenwyk (1979), Pest Management Guide for Insects and Nematodes of Cotton in California, Division of Agricultural Sciences, University of California, Berkeley, CA, 62 pp.

Watt, K. E. F. (1963), "Dynamic programming, 'look ahead programming', and the strategy of insect pest control", The Canadian Entomologist 95: 525-536.

Watt, K. E. F. (1964), "The use of mathematics and computers to determine optimal strategy and tactics for a given insect pest control problem", The Canadian Entomologist 96: 202-220.

Wickwire, K. (1977), "Mathematical models for the control of pests and infectious diseases: a survey", Theoretical Population Biology 11: 182-238.

Wilson, L. T. and W. W. Barnett (1983), "Degree-days: an aid in crop and pest management, California Agriculture 37: 4-7.

DEPARTMENT OF MATHEMATICS
UNIVERSITY OF CALIFORNIA
DAVIS, CALIFORNIA 95616

ECONOMIC INCENTIVES FOR POLLUTION CONTROL

Maureen L. Cropper

1. INTRODUCTION. All economies must by some means determine what resources to devote to pollution control. When the number of polluters and victims is small this decision is often reached through a bargaining process. Bargaining, however, becomes costly as the number of parties increases. Pollution problems involving many polluters or victims are therefore solved by a neutral party, the government, which imposes penalties on polluters to minimize the total costs of pollution.

In either case determining the appropriate level of pollution requires three types of information: the costs to firms of reducing emissions, the damages to victims associated with ambient pollution, and the relationship between emissions and ambient pollution. Given this information, and assuming that emissions are costlessly observable, it is a simple matter to design an optimal contract in the bargaining case, or an optimal regulatory scheme when the government controls pollution, which is a function of emissions.

What makes the pollution control problem difficult is that the necessary information is often known imperfectly. The purpose of these notes is to examine how incentives for pollution control should be structured depending on what is known about damages, abatement costs, and emissions.

Section 2 of the notes models agreements between a single polluter and a single victim, either one of whom may initially be assigned rights to the environment. When abatement costs, damages, and emissions are known to both parties one can design a contract which improves the welfare of at least one party (a Pareto optimal contract) regardless of the initial assignment of property rights. In many situations, however, emissions can be monitored by the victim only at great cost. If the relationship between emissions and damages is also known imperfectly, incentives cannot be based on emission levels. One question in this case is whether the first-best solution to the problem—the solution which would be reached if emissions were costlessly observable—can be reached with imperfect information. When a first-best solution cannot be reached the best solution achievable can be characterized.

Section 3 considers many-firm pollution problems, which are typically solved by the government imposing penalties on polluters. Here a first-best solution may also be unachievable due to the high cost of monitoring emissions, and the questions addressed in Section 2 arise again. A different information problem, however, becomes important in the many-firm case. Since the government must regulate thousands of firms in diverse industries it is expensive to obtain information about all firms' abatement costs. The government can, of course, request that firms provide this information; however, if firms suspect that the information will be used to set pollution taxes they will have an incentive to misrepresent costs. A natural question is whether a system of taxes can be devised which will induce firms to correctly reveal their cost functions. The answer, under certain conditions, is yes; however, the cost of communication required to elicit this information may be prohibitive. In this case one can view the government's ignorance about costs as a constraint and characterize the best solution achievable given this information.

2. OPTIMAL POLLUTION REGULATION VIA BARGAINING. The two-party pollution problem may be illustrated by two firms, a farm and a papermill, located on a river. The papermill, located upstream from the farm, wishes to use the river to dispose of organic waste, while the farm wishes to use the river for irrigation. Suppose that emissions e, $e \in R_+^1$, are a byproduct of an m-dimensional vector of outputs $\underline{y}$, $\underline{y} \in R_+^m$, produced by the papermill. Let $G(\underline{y},e)$ denote the minimum total cost of producing a given $(\underline{y},e)$ vector and let $H(\underline{y})$ denote the revenue which the firm receives from a given output vector. The cost to the firm of emitting e of pollution is defined to be the difference between maximum profits if emissions are unconstrained and maximum profits if emissions are e,

$$C(e) \equiv \max_{\underline{y},e}[H(\underline{y})-G(\underline{y},e)] - \max_{\underline{y}}[H(\underline{y})-G(\underline{y},e)]. \tag{1}$$

We assume that H is a concave function of $\underline{y}$ and that G is a strictly convex function of $(\underline{y},e)$ so that $C(e)$ is a strictly convex function of e. We also assume that $C'(e) \leq 0$ for $0 \leq e \leq \bar{e}$, where $\bar{e}$ is the level of emissions which the papermill would choose if unconstrained, i.e., $C(\bar{e}) = 0$. Since $C(e)$ is the cost of reducing emissions from $\bar{e}$ to e, C will sometimes be referred to as the cost of pollution abatement rather than the cost of emissions.

Damages to the farm from water pollution depend not on emissions themselves but on ambient water quality. Let x, $x \in R_+^1$, represent ambient water pollution and define damages D associated with x as the difference between the farm's profits when pollution is x and profits in the absence of pollution,

$$D = D(x), \quad D'(x) > 0, \quad D''(x) > 0. \tag{2}$$

2.1. BARGAINING UNDER PERFECT INFORMATION. Suppose in this example that the cost and damage functions are known to both firms, and that ambient pollution is a strictly increasing, convex function of e, $x = x(e)$, which is also known by both firms. If e is costlessly observable by both parties, or if x is observable so that e can be inferred from $x(e)$, then it is easy to construct an optimal pollution contract, i.e., one that makes at least one party better off compared with the pre-contract situation.

Let $C_0 \equiv C(e_0)$ denote abatement costs paid by the polluter in the pre-contract situation and $D_0 \equiv D(x(e_0))$ denote damages initially suffered by the victim. If at some $0 \leq e \leq \bar{e}$, $e \neq e_0$ the net change in abatement costs plus damages is positive,

$$C_0 - C(e) + D_0 - D(x(e)) > 0, \quad 0 \leq e \leq \bar{e}, \tag{3}$$

opportunities exist for a mutually beneficial agreement. Let $S(e)$ denote the amount which the victim pays the polluter to set emissions at e. (If $S(e) < 0$ then the polluter pays the victim.) A Pareto optimal contract is a function S which maximizes the bargaining gains to one party, e.g., the victim, subject to the constraint that the other party be no worse off than in the pre-contract situation. Formally, $S(e)$ must satisfy

$$\max_S D_0 - D(x(e)) - S(e)$$
$$\text{s.t.} \quad S(e) - C(e) \geq -C_0 \quad \text{and} \quad e = \operatorname{argmax}[S(e)-C(e)]. \tag{4}$$

It is easily verified that (5) constitutes a Pareto optimal contract,

$$S(e) = D_0 - D(x(e)) - k,$$
$$k = D_0 - D(e^*) + C_0 - C(e^*), \tag{5}$$

where e^* is the value of e that maximizes the left-hand side of (3). Equation (5) induces the polluter to set emissions at the level that minimizes damages plus abatement costs and divides net benefits according to initial bargaining positions.

For (5) to be enforceable, however, both parties must know the functions $D(x)$, $C(e)$, and $x(e)$ and must be able to observe x or e. These conditions will not be met if emissions are impossible to monitor, as in the case of agricultural runoff. Inability to monitor e presents no problem if x is cheaply observable and $x(e)$ known; however, the relationship between emissions and ambient pollution may be uncertain as well. This is precisely the situation which would obtain if the polluter were a farm and the victim a fisherman. Section 2.2 defines and characterizes a Pareto optimal contract under these circumstances.

2.2. BARGAINING UNDER IMPERFECT INFORMATION. To focus on the problem of monitoring emissions suppose that both firms know $C(e)$ and $D(x)$ but regard the relationship between ambient pollution and emissions as uncertain. Formally, ambient pollution is a function of emissions and a random variable θ, $\theta \in R^1$, where θ might represent water temperature or stream flow,

$$x = x(e,\theta), \quad \partial x/\partial e > 0, \quad \partial^2 x/\partial e^2 > 0, \quad \text{all} \quad \theta. \tag{6}$$

At the time the contract is negotiated both firms have identical probability distributions on θ. After the contract is negotiated the realized value of θ may be observed by one or both parties. In Model 1 the polluter remains uncertain about θ when e is chosen; in Model 2 the realized value of θ is known by him before he chooses e. In either case the realized value of θ may be observable by both parties after x occurs so that a contract can be based on θ as well as x.

We now consider the form which pollution contracts may take. Let $\underline{z}$ denote the variables observable by both parties after e has been chosen and θ realized. In this section $\underline{z}$ cannot include e but includes x and may include the realized value of θ. Let $S(\underline{z})$ be the payment made by the victim to the polluter. A Pareto optimal contract is defined as a function $S(\underline{z})$ which maximizes expected utility of benefits to the victim subject to the constraint that the expected utility of benefits to the polluter not fall below their pre-contract level, and to the constraint that the value of emissions chosen maximizes the expected utility of the polluter. Formally, the function $S(\underline{z})$ must satisfy (7)-(9),

$$\max_{S} E_\theta U^V[D_0 - D(x(e,\theta))-S(\underline{z})] \equiv V^V(S(\underline{z}),e) \tag{7}$$

$$\text{s.t.} \quad E_\theta U^P[S(\underline{z})-C(e)] \geq U^P(-C_0) \tag{8}$$

$$\text{and} \quad \max_{e} E_\theta U^P[S(\underline{z})-C(e)] \equiv V^P(S(\underline{z}),e) \quad \text{(Model 1)} \tag{9a}$$

$$\text{or} \quad \max_{e} E_\theta U^P[S(\underline{z})-C(e(\theta))] \equiv V^P(S(\underline{z}),e) \quad \text{(Model 2)}. \tag{9b}$$

U^V and U^P, the utility functions of the victim and the polluter, are assumed to be strictly increasing and concave. Equations (9a) and (9b) are assumed to have unique solutions.

This definition of Pareto optimality differs in two respects from (4). Equation (4) treats the functions U^v and U^p as linear, implying that

$$E_\theta U[D(\theta)] = U[E_\theta(D(\theta))], \tag{10}$$

i.e., that both firms are risk-neutral. If (10) holds then neither firm requires compensation for uncertainty about pollution, a reasonable assumption if each firm is owned by a large number of investors with diversified portfolios. If, however, a firm is closely-held and if uncertainty about pollution damages is large it may be reasonable to assume that the firm is risk-averse,

$$E_\theta U[D(\theta)] < U[E_\theta(D(\theta))], \tag{11}$$

and requires compensation for uncertainty regarding the outcome of the bargaining agreement. In this case a Pareto optimal contract must share this risk between the polluter and victim, as well as providing incentives for the polluter to alter his emissions.

Secondly, since the pollution contract cannot be based on emissions, equation (9a) or (9b) may be binding. A first-best contract is defined as one in which e and S are chosen to maximize (7) subject only to (8), i.e., a contract which could be realized if e were observable.

Given these definitions there are two questions of interest: (1) When emissions are not observable can a first-best contract be achieved? (2) If such a solution cannot be achieved, what does a Pareto optimal contract look like? Answers to both questions are provided by the literature on the principal-agent problem (Harris and Raviv [4], Holmstrom [5], Shavell [8]), of which the pollution problem is a specific example.

As shown by Harris and Raviv [4], there are two conditions under which a first-best solution is achievable when emissions cannot be observed. One is when the realized value of θ is observable by both parties ex post so that the contract can be based on θ as well as on x. The other occurs when θ is not observable by both parties but the polluter is risk-neutral. These results, which hold for models 1 and 2, are stated formally as

PROPOSITION 1. Any contract based on $\underline{z} = (x \theta e)$ can be dominated by a contract based on $\underline{z} = (x \theta)$ in the sense that $V^v(S(\underline{z}),e)$ and $V^p(S(\underline{z}),e)$ are at least as great for both parties when $\underline{z} = (x \theta)$.

PROPOSITION 2. If the polluter is risk-neutral any contract based on $\underline{z} = (x \theta e)$ can be dominated by a contract based on $z = x$.

An important implication of these propositions is that monitoring emissions is of no value when θ is observable or the polluter is risk-neutral.

In the farm-fisherman example a first-best contract is achievable as long as ambient water quality can be cheaply monitored and weather conditions influencing water quality, assuming they are the source of θ, are also observable. The other case in which monitoring of emissions is unnecessary is when the polluter maximizes expected profits. In this case there is a Pareto optimal contract (Shavell [8, Prop. 4]) which pays the polluter $D_0 - D - k$ and the victim k. This solves the incentive problem by forcing the polluter to consider the effect of his emissions on the victim and provides for optimal risk sharing, i.e., the risk-neutral polluter bears all the risk.

When the polluter is risk-averse and θ is not observable ex post the Pareto optimal contract is no longer a first-best contract and there are gains to monitoring emissions. Since perfect monitoring is by assumption too costly, it is natural to ask whether an optimal contract can be improved upon by imperfect monitoring. The victim, for example, may be unable to observe emissions but may have information about a related variable, y, e.g. stream color. Suppose that y is a function of e and a random variable δ. Should the pollution contract be based on y given that this increases uncertainty about the payoff and given that the polluter is risk-averse? For Model 1 Shavell [8, Prop. 5] has shown that as long as the distribution of y depends on e, a contract based on x and y exists which dominates a contract based solely on x. In the special case in which $y = e + \delta$, where δ is distributed independently of θ on the interval $[\delta_0, \delta_1]$, $\delta_0 < 0 < \delta_1$, Harris and Raviv [4] demonstrate that the optimal contract is dichotomous,

$$S(x,y) = \begin{cases} S_0(x,y) & \text{if } y \leq \hat{y} \\ w & \text{if } y > \hat{y} \end{cases}, \quad (12)$$

i.e., a payment which is a continuous function of ambient pollution and output is received if output falls below a critical level and a constant amount is received if output exceeds that level. This implies, in particular, that if e is costlessly observable a first-best solution can always be attained by means of a dichotomous (forcing) contract.

3. POLLUTION CONTROL WHEN BARGAINING IS NOT POSSIBLE. When the number of polluters and/or victims is large, voluntary agreements of the type described in Section 2 are too costly to arrange and pollution is usually regulated by a neutral third party, the government. Even under full information this changes the structure of the pollution problem considerably. Pollution regulation is no longer a voluntary agreement between two parties but a set of penalties imposed on firms which may reduce their utility below that received in the

absence of regulation. The definition of an optimal set of penalties is, however, similar to the definition of a Pareto optimal contract in the bargaining case. In Section 2 a necessary condition for a contract to be Pareto optimal is that it minimizes $D(x) + C(e)$. In the multi-firm case an optimal set of penalties is that which minimizes the sum of control costs plus damages.

As in the two-person problem, it is convenient to describe the solution to the multi-firm pollution problem under full information. Section 3.2 considers how this solution is altered when emissions cannot be observed, while Section 3.3 focuses on the government's ignorance of firms' cost functions.

3.1. POLLUTION CONTROL UNDER FULL INFORMATION. Consider n firms located in the same geographic area, each of which emits an amount e_i, $i = 1,\ldots,n$, of a pollutant such as particulate matter or sulfur dioxide. The ambient level of the pollutant, x, is a known, convex function of the vector $\underline{e} = (e_1 e_2 \cdots e_n)$, $x(\underline{e})$. Let x_i denote the partial derivative of this function with respect to e_i and assume $x_i > 0$, $i = 1,\ldots,n$. As in the bargaining case damages to society are assumed to be an increasing, strictly convex function of ambient pollution, $D = D(x)$. The cost of emissions e_i to firm i is given by $C(e_i,\alpha_i)$, where C is defined by equation (1). α_i, which could be a vector but for notational simplicity is treated as a scalar, is a parameter which distinguishes firm i's abatement cost function from firm j's. As in Section 2 C is assumed to be a strictly convex function of e_i for each value of α_i.

Given the functions C, D and x and the parameter vector $\underline{\alpha} = (\alpha_1 \cdots \alpha_n)$ it is a simple matter to determine the vector of emissions $\underline{e}^*$ which minimizes the sum of damages plus abatement costs,

$$\underline{e}^* = \operatorname{argmin}[D(x(\underline{e})) + \sum_{i=1}^{n} C(e_i,\alpha_i)]. \qquad (13)$$

Previous assumptions guarantee that the solution to (13) is unique and is characterized by the necessary conditions

$$D'(x(\underline{e}^*))x_i(\underline{e}^*) = -C_1(e_i^*,\alpha_i) \quad i = 1,\ldots,n, \qquad (14)$$

where $C_1(e_i,\alpha_i) \equiv \partial C/\partial e_i$.

Assuming that each firm chooses e_i to minimize the sum of abatement costs plus taxes, and that the government can costlessly observe emissions, the full-information solution to the pollution control problem can be achieved by levying on firm i a tax t_i per unit of emissions,

$$t_i = D'(x(\underline{e}^*))x_i(\underline{e}^*). \qquad (15)$$

3.2. POLLUTION CONTROL WITH COSTLY MONITORING. As in the two-party case, the cost of monitoring emissions may be so great in the multi-firm case that one wishes to consider only the class of penalties based on ambient pollution levels. Since this topic has been treated in detail in Section 2 we emphasize ways in which the results of that section are altered by considering multiple firms.

In Section 2 the victim's inability to observe $\underline{e}$ prevents a first-best solution from being attained only if the relationship between ambient pollution and emissions is uncertain. With many firms removing this uncertainty does not solve the monitoring problem since any one firm's emissions cannot be inferred by observing x even if the function $x(\underline{e})$ is known. We therefore ask whether the full-information solution can be attained through penalties which depend only on x when $x(\underline{e})$ is known.

The answer, as shown by Holmstrom [6], is yes. Define T_i, the penalty on the ith firm, as

$$T_i = 0 \quad \text{if} \quad x \leq x(\underline{e}^*)$$
$$T_i = b_i \quad \text{if} \quad x > x(\underline{e}^*) \tag{16}$$

where $b_i > C(e_i^*, \alpha_i)$. It is easily verified that firm i minimizes $T_i + C(e_i, \alpha_i)$ by setting $e_i = e_i^*$, assuming that $e_j = e_j^*$ for all $j \neq i$. In other words $\underline{e}^*$ is a Nash (non-cooperative) equilibrium if penalties are set according to (16).

A similar solution is obtainable when all polluters are risk-neutral and the relationship between ambient pollution and emissions is uncertain. Let $F(\bar{x}, \underline{e})$ be the probability that $x \geq \bar{x}$ given that emissions are $\underline{e}$, and assume that $\partial F/\partial e_i \equiv F_i(\bar{x}, \underline{e}) > 0$, all i, and that $F(\bar{x}, \underline{e})$ is a strictly convex function of $\underline{e}$. If penalties of the form

$$T_i = 0 \quad \text{if} \quad x \leq \bar{x}$$
$$T_i = d_i \quad \text{if} \quad x > \bar{x} \tag{17}$$

are imposed on firm i the firm will set

$$F_i(\bar{x}, \underline{e})d_i + C_1(e_i, \alpha_i) = 0 \tag{18}$$

to minimize the expected value of abatement costs plus penalties. By choosing

$$d_i = -C_1(\hat{e}_i, \alpha_i)/F_i(\bar{x}, \hat{\underline{e}}) \tag{19}$$

one guarantees that firm i sets $e_i = \hat{e}_i$ provided that all firms $j \neq i$ set $e_j = \hat{e}_j$; i.e., one guarantees that $\hat{\underline{e}}$ is a Nash equilibrium. A penalty of the form (17) can thus be used to achieve the emissions vector that minimizes expected damages plus abatement costs when $\underline{e}$ is unobservable and the relationship between ambient pollution and emissions is uncertain.

3.3. POLLUTION CONTROL UNDER IMPERFECT INFORMATION ABOUT FIRM TECHNOLOGY. To focus on the problem of monitoring emissions it has been assumed that control cost and damage functions are known by all parties. Although it is reasonable to assume that each firm knows its cost of emissions function, it is unlikely that the government knows $\underline{\alpha}$ with certainty. The government is also unlikely to know the damage function $D(x)$; however, the two information problems are fundamentally different. Since ignorance of $\underline{\alpha}$ is a problem of asymmetrical information one can ask whether the government can induce firms to truthfully report $\underline{\alpha}$ and thus achieve the full-information solution $\underline{e}^*$. Ignorance about $D(x)$, however, more likely represents scientific ignorance, e.g., ignorance about the health effects of pollution, rather than an information asymmetry. It is therefore natural to redefine the optimal pollution vector taking uncertainty about $D(x)$ as given. Since this will not alter the structure of optimal pollution penalties we concentrate instead on the problem of asymmetric information regarding firms' costs. To focus on this problem it is assumed that emissions are costlessly observable by all parties and that the $x(\underline{e})$ and $D(x)$ functions are known.

We begin by noting that the government cannot simply request that each firm report its α_i, for if the firm suspects that this information will be used to compute emissions taxes it will misreveal α. Specifically, if firms believe that taxes will be set according to (15) and if all firms affect ambient pollution symmetrically,

$$x = x(\sum_{i=1}^{n} e_i), \qquad (20)$$

it can be shown that each firm has an incentive to understate its marginal cost of pollution abatement (Kwerel [7]).

To show this, suppose that α_i increases the marginal cost of pollution abatement,

$$-\partial^2 C/\partial e_i \partial \alpha_i > 0, \quad \text{all} \quad \alpha_i. \qquad (21)$$

Let $\hat{\alpha}_i$ denote the value of α_i revealed to the government (α_i denotes the true parameter value), and $\hat{\underline{\alpha}}_{-i}$ denote the vector of parameters revealed by firms other than i. $\hat{e}_i$ denotes the value of emissions which minimizes firm i's taxes plus abatement costs, $t_i e_i + C(e_i, \alpha_i)$, if $\hat{\underline{\alpha}}$ is reported to the government. Since

$$\partial[t_i\hat{e}_i + C(\hat{e}_i,\alpha_i)]/\partial\hat{\alpha}_i = (\partial t_i/\partial\hat{\alpha}_i)\hat{e}_i \qquad (22)$$

the firm clearly has an incentive to set $\hat{\alpha}_i < \alpha_i$ if t_i is an increasing function of $\hat{\alpha}_i$ for any $\hat{\underline{\alpha}}_{-i}$.

To see that $\hat{\alpha}_i$ increases t_i note that when (20) holds (14) reduces to

$$D'(\sum \hat{e}_i^*) = -C_1(\hat{e}_i^*,\hat{\alpha}_i) \quad i = 1,\ldots,n, \qquad (23)$$

where $\hat{\underline{e}}^*$ denotes the optimal emissions vector given that firms reveal $\hat{\underline{\alpha}}$. In this case all firms optimally produce at the same marginal emissions cost and tax rates facing all firms are identical,

$$t_i = t = D'(\sum \hat{e}_i^*), \text{ all } i. \qquad (24)$$

That $\hat{\alpha}_i$ raises t follows from (23) and (24) which together imply

$$\frac{\partial t}{\partial \hat{\alpha}_i} = D'' \sum_{j=1}^{n} \frac{\partial \hat{e}_j}{\partial \hat{\alpha}_i} = -D'' \frac{\partial^2 C/\partial e_i \partial \hat{\alpha}_i}{\partial^2 C/\partial e_i^2 + D''} > 0. \qquad (25)$$

In view of this result the question is whether there exist taxes different in form from (15) which will induce firms to reveal their true α_i's. An affirmative answer to this question has been provided by Groves and Loeb [3]. To define the tax levied on firm i recall that $\hat{e}_j^*$ denotes optimal emissions for firm j given that firms report $\hat{\underline{\alpha}}$ to the government. To emphasize that $\hat{e}_j^*$ is a function of firm i's reported α, $\hat{\alpha}_i$, as well as of the values of α reported by all other firms, write

$$\hat{e}_j^* = \hat{e}_j^*(\hat{\alpha}_i, \hat{\underline{\alpha}}_{-i}). \qquad (26)$$

The Groves-Loeb tax paid by firm i, which depends both on $\hat{\alpha}_i$ and on e_i, is defined as

$$T_i(e_i,\hat{\alpha}_i,\hat{\underline{\alpha}}_{-i}) = D(x(e_i,\hat{\underline{e}}_{-i}^*(\hat{\alpha}_i,\hat{\underline{\alpha}}_{-i}))) \\ + \sum_{j\neq i} C(\hat{e}_j^*(\hat{\alpha}_i,\hat{\underline{\alpha}}_{-i}),\hat{\alpha}_j) + A_i(\hat{\underline{\alpha}}_{-i}) \qquad (27)$$

where $\hat{\underline{e}}_{-i}^*$ denotes the optimal vector of emissions for all firms other than i and A_i is a firm-specific function.

For the tax functions (27) to induce truthful reporting of α_i regardless of the parameters reported by other firms it must be the case that

$$C(\hat{e}_i(\alpha_i, \hat{\underline{\alpha}}_{-i}),\alpha_i) + T_i(\hat{e}_i(\alpha_i,\hat{\underline{\alpha}}_{-i}),\alpha_i,\hat{\underline{\alpha}}_{-i}) \\ \leq C(\hat{e}_i,\alpha_i) + T_i(\hat{e}_i,\hat{\alpha}_i,\hat{\underline{\alpha}}_{-i}) \quad \text{all } \hat{\alpha}_i \neq \alpha_i. \qquad (28)$$

To see that this is so, note that from the definition of $\hat{e}_j^*$ and T_i

$$\begin{aligned}
C(\hat{e}_i(\alpha_i,\hat{\underline{\alpha}}_{-i}),\alpha_i) &+ T_i(\hat{e}_i(\alpha_i,\hat{\underline{\alpha}}_{-i}),\alpha_i,\hat{\underline{\alpha}}_{-i}) - A_i(\hat{\underline{\alpha}}_{-i}) \\
&= \min_{\underline{e}_{-i}}[D(e_i(\alpha_i,\hat{\underline{\alpha}}_{-i}),\underline{e}_{-i}(\alpha_i,\hat{\underline{\alpha}}_{-i})) \\
&\qquad + \sum_{j \neq i} C(e_j,\hat{\alpha}_j) + C(e_i,\alpha_i)] \\
&\leq D(e_i(\hat{\alpha}_i,\hat{\underline{\alpha}}_{-i}),\underline{e}_{-i}(\hat{\alpha}_i,\hat{\underline{\alpha}}_{-i})) \\
&\qquad + \sum_{j \neq i} C(e_j,\hat{\alpha}_j) + C(e_i,\alpha_i) \\
&= C(\hat{e}_i,\alpha_i) + T_i(\hat{e}_i,\hat{\alpha}_i,\hat{\underline{\alpha}}_{-i}) - A_i(\hat{\underline{\alpha}}_{-i}).
\end{aligned} \qquad (29)$$

Thus for any $\hat{\underline{\alpha}}_{-i}$, firm i's dominant strategy is to report α_i.

There are, however, two drawbacks to the Groves-Loeb tax. Under the tax it is optimal for firm i to report its true α_i only if the firm cannot communicate with other firms. There is in general no tax scheme which will induce true revelation of information if firms can form coalitions (Green and Laffont [2]).

Secondly, for the government to impose a tax which achieves the first-best solution two rounds of communication are required. In round one the government sends firms tax functions (27) and receives the parameter vector $\underline{\alpha}$. During round two the $\underline{\alpha}$ vector is communicated to each firm, and taxes are recomputed with $\hat{\underline{\alpha}}_{-i}$ replaced by $\underline{\alpha}_{-i}$. If such communication is not possible then the first-best solution cannot be achieved (Dasgupta, Hammond and Maskin [1]).

This implies that one must adopt an alternative definition of an optimal incentive structure if the government must announce taxes based only on its initial information. To formalize what is known about $\underline{\alpha}$ suppose that all firms and the government view the $\{\alpha_i\}$ as independent drawings from a publicly known probability distribution. The realized value of α_i is known only to firm i. In defining the optimal set of taxes suppose that firm i can be taxed only on its own emissions, e_i, since to tax the firm based on $\underline{e}_{-i}$ would subject it to uncertainty regarding other firms' costs. Let $\tilde{e}_i$ denote firm i's optimal response to the tax function $T_i(e_i)$, i.e.,

$$\tilde{e}_i = \text{argmin } T_i(e_i) + C_i(e_i,\alpha_i). \qquad (30)$$

The optimal tax functions $T_i(e_i)$ are those which minimize the sum of expected damages plus pollution costs,

$$E[D(\tilde{e}_1(T_1(\ ),\alpha_1),\ldots,\tilde{e}_n(T_n(\ ),\alpha_n)) + \qquad (31)$$

$$\sum_{i=1}^{n} C_i(\tilde{e}_i(T_i(\ ),\alpha_i),\alpha_i)],$$

given firms' response functions.

In the case in which the marginal damage function is linear in $\underline{e}$ and each marginal emission cost function is linear in e_i and α_i Weitzman [9] has shown that the optimal tax is a combination of a per-unit tax on emissions, t_i, and a penalty for deviating from the emissions quota e_i^*,

$$T(e_i) = t_i e_i + q_i (e_i - e_i^*)^2. \qquad (32)$$

The emissions quotas $\{e_i^*\}$ are those levels of e_i that minimize the sum of expected damages plus control costs,

$$\underline{e}^* = \text{argmin } E[D(\underline{e}) + \sum_{i=1}^{n} C_i(e_i,\alpha_i)]. \qquad (33)$$

t_i is the marginal damage caused by firm i when $\underline{e} = \underline{e}^*$,

$$t_i = \partial D(\underline{e})/\partial e_i \Big|_{\underline{e}=\underline{e}^*}, \quad i = 1,\ldots,n. \qquad (34)$$

The cost of the firm of deviating from the expected-loss-minimizing emissions level depends on the slope of the marginal damage function. Since $\partial D/\partial e_i$ is assumed linear in $\underline{e}$ we may write

$$\partial D/\partial e_i = t_i - \sum_{j=1}^{n} \beta_{ji}(e_j - e_j^*), \quad i = 1,\ldots,n, \qquad (35)$$

where t_i is defined by (34). q_i varies directly with β_{ii}, the slope of the marginal damage function with respect to firm i's emissions,

$$q_i = \beta_{ii}/2. \qquad (36)$$

Under the foregoing assumptions the optimal taxation scheme has an intuitively appealing interpretation. When marginal damages increase slowly with emissions the optimal penalty is a linear tax on emissions. Although it is uncertain how much this tax will reduce emissions, the cost of deviating from the optimal emission level is small since damages increase slowly with emissions. When the marginal damage function is steep, however, deviations from $\underline{e}^*$ are costly to society and the firm is therefore penalized heavily for deviating from the optimal emissions level.

BIBLIOGRAPHY

[1] P. Dasgupta, P. Hammond and E. Maskin, On imperfect information and optimal pollution control, Rev. Econ. Studies, 47 (1980), 857-860.

[2] J. R. Green and J.-J. Laffont, Incentives in Public Decision-Making, North Holland, Amsterdam, 1979.

[3] T. Groves and M. Loeb, Incentives and public inputs, J. Pub. Econ., 4 (1975), 211-226.

[4] M. Harris and A. Raviv, Optimal incentive contracts with imperfect information, J. Econ. Theory, 20 (1979), 231-259.

[5] B. Holmstrom, Moral hazard and observability, Bell J. Econ., 10 (1979), 74-91.

[6] B. Holmstrom, Moral hazard in teams, Bell J. Econ., 13 (1982), 324-340.

[7] E. Kwerel, To tell the truth: imperfect information and optimal pollution control, Rev. Econ. Studies, 44 (1977), 595-601.

[8] S. Shavell, Risk sharing and incentives in the principal and agent relationship, Bell J. Econ., 10 (1979), 55-73.

[9] M. Weitzman, Optimal rewards for economic regulation, Amer. Econ. Rev., 68 (1978), 683-691.

DEPARTMENT OF ECONOMICS
UNIVERSITY OF MARYLAND
COLLEGE PARK, MARYLAND 20742

DEPLETION AND DISCOUNTING:
A CLASSICAL ISSUE IN THE ECONOMICS OF EXHAUSTIBLE RESOURCES

GEOFFREY HEAL[1]

1. INTRODUCTION

In the discussion of intertemporal resource-allocation, a central place has always been occupied by the issue of striking an appropriate balance between the demands of the present generation and those of their successors. In this context, the practice of discounting and the selection of a discount rate have loomed large. My paper addresses the practice of discounting, and the technical mathematical role played by this practice.

One can naturally pose the problem of selecting a depletion program for an exhaustible resource as a constrained maximization problem. I shall show below that discounting is "nearly necessary" if the maximand in this problem is to be continuous: a failure to give less weight to benefits from resource use that accrue in the future than is given to those accruing currently, generates a fundamental discontinuity in the choice problem associated with selecting a depletion program. There is therefore a choice: give less weight to future benefits than to present ones, and have a well-posed problem, or give equal weight and face a potentially ill-posed and insoluble problem.

For those determined to place equal value on the interests of present and future generations, this may seem a Faustian choice. I shall in fact suggest that by suitable selection of parameters other than the discount rate, it can be avoided. First, however, I set up the problem.

Consider an economy with a fixed stock of S_0 units of an exhaustible resource. In each of a countable sequence of time periods the economy consumes this resource, using up c_t units in period t. Economically it makes sense that $c_t \geq 0$ for all t, and the exhaustibility of the resource implies that necessarily $\sum_t c_t \geq S_0$. Let $c = (c_1, c_2, \ldots, c_t, \ldots)$, and define

$$F = \{c : c_t \geq 0 \; \forall \; t \text{ and } \sum_t c_t \leq S_0\} .$$

---

[1] I am grateful to William Brock, Graciela Chichilnisky and William Zame for helpful comments on this problem.

Then F is the set of feasible consumption sequences.

A very natural economic question to ask is: "At what rate should we deplete our fixed stock S_0 of this resource?" More formally, one might ask for the best feasible consumption sequence, which of course requires us to define an order on F and pick a maximal element.

This provides the simplest possible framework for studying optimal depletion policies: picking an element of F maximal under some appropriate ordering. To complete this research program, one needs to derive an ordering from underlying economic considerations, to show that there is a maximal element of F, and to characterize this element. Of these three steps, only the third is straightforward; deriving and justifying an ordering, and demonstrating that there is an element of F maximal under this, are both surprisingly difficult tasks.

In the next section I shall study these two issues -- ordering F, and finding a best element. I shall show that our ability to solve these problems depends on our willingness to make certain value judgments about the relative merits of present and future consumption, and in particular depends on our willingness to discount benefits from future consumption relative to benefits from present consumption. Section 3 will discuss the characterization of optimal depletion policies, and Section 4 will discuss extensions of the basic model to a more general framework incorporating the possibility of future resource discoveries or of resource-substituting technical change, both occurring in an unpredictable (stochastic) fashion.

2. OPTIMAL DEPLETION?

I want to start by investigating the consequences of using a very simple and wide-spread method of generating an ordering over F. Suppose that consuming c_t units of the resource in period t yields an economic benefit $B_t(c_t)$, and assume furthermore that $B_t(c_t)$ is a continuous increasing and strictly concave real-valued function. At this level of abstraction, we need not be concerned with the units in which B_t is measured, nor with how we evaluate $B_t(c_t)$. The assumptions that $B_t(\ )$ is increasing and strictly concave imply that the benefits from resource use increase with the level of resource use, but do so at a diminishing rate. Economists would invoke the classical (but nevertheless questionable) assumption of diminishing returns to justify this. The total benefit associated with a consumption profile $c = (c_1, c_2, \ldots, c_t, \ldots)$ is $\sum_t B_t(c_t)$, assuming for the moment that this is defined. I want to investigate the consequences of using this total to rank the elements of F: define an ordering p on F by, $c_1 \in F$, $c_2 \in F$,

$$c_1 p c_2 \iff \sum_t B_t(c_{1t}) \geq \sum_t B_t(c_{2t}) .$$

Even when the infinite sums are well-defined, this approach can lead to difficulties. To see this, let $B_t(c_t) = B(c_t)$ for all t, and take the number of time periods to be finite and equal to T. Then we rank elements of F by $\sum_{t=1}^{T} B(c_t)$. Obviously, there exists in this case a best element of F, given by $c_t = S_0/T$, $t = 1, \ldots, T$. Look at the problem "maximize $\sum_{t=1}^{T} B(c_t)$, subject to $c \in F$." It is easy to verify that for T finite or infinite, a necessary condition for a solution is that $c_t = c$ for all t. This is also sufficient if $\sum_t c_t = S_0$, giving $c_t = S_0/T$.

Now check what happens as $T \to \infty$. Our chosen consumption path tends uniformly to zero. Obviously $c_t = 0 \ \forall \ t$ is not the best possible path, yet it is the only constant path in F, and any path on which c_t is not constant violates the necessary conditions for a solution to the maximization problem.

So the infinite-horizon version of our problem has no solution: in the infinite-horizon case there is no maximal element of F under the ordering p. In economic terms, does this matter? Unfortunately, the answer is generally believed to be "yes." The reason is that in a long-run planning problem of this type, the choice of any finite horizon T involves the designation of a date beyond which the benefits of resource use are no longer to be counted. An optimal depletion policy with respect to an ordering defined only on the first T periods will certainly exhaust the entire resource stock by the end of period T. Few people are sufficiently certain of their apocalyptic visions to make this choice of T with confidence, and the accepted convention has been to avoid the problem and make T infinite. There is also, I think, a general unease about taking seriously finite-horizon solutions whose limit with the time horizon is something as unappealing as the zero-consumption path. (For a more detailed discussion, see [11].)

It follows, then, that we would like to reformulate our problem so that it has solutions in both its finite and infinite horizon forms. To do this, we have to find conditions sufficient to ensure that there is a solution to the infinite-horizon case. Recall the problem:

"maximize $V = \sum_{t=1}^{\infty} B_t(c_t)$ subject to

$$(c_1, c_2, \ldots, c_t, \ldots) \in \{c : c_t \geq 0, \sum_{t=1}^{\infty} c_t \leq S_0\} = F" . \tag{A}$$

This has the structure of maximizing a real-valued function on a set F. F is a set of summable sequences, and we shall consider it to be a subset of the of the Hilbert space ℓ_2 of sequences $c = (c_1, c_2, \ldots)$ for which $\sum_t (c_t)^2 < \infty$, with the inner product $\langle x, y \rangle = \sum_t x_t y_t$. Clearly F is a bounded, convex set. We have assumed the B_t to be continuous from R^1 to R^1, increasing and strictly concave: hence V is increasing and strictly concave.

THEOREM 1. *If V is continuous in the ℓ_2 norm, then there exists a solution to problem* (A), *i.e., there exists an optimal depletion profile for the infinite-horizon case.*

PROOF: The set F is clearly bounded in the ℓ_2 norm, and is convex. It is also ℓ_2-closed. To see this, consider a sequence f^n of points F, with $f_n \xrightarrow{\ell_2} f$. We have to show that $f \in F$. Now f^n converging in the ℓ_2 norm to f implies that f^n converge pointwise to f. So for any $\varepsilon > 0 \ \exists \ n(\varepsilon)$ such that for $n \geq n(\varepsilon)$, $\sum_{t=1}^{\infty} (f_t^n - f_t) < \varepsilon$. It follows that $\sum_{t=1}^{\infty} f_t \leq S_0$, i.e., F is ℓ_2-closed.

We have an ℓ_2-continuous concave real-valued function V defined on a convex ℓ_2-bounded ℓ_2-closed set F. The proof now follows that of Theorem 1 of Chichilnisky and Kalman [6] or Lemma 6 of Chichilnisky and Heal [4]. By Alaoglu's Theorem [10], F is weakly compact. Define

$$S = \sup_{c \in F} V(c).$$

Consider a sequence $\{c^n\} \in F$, with $\ldots V(c^{n+1}) > V(c^n) > V(c^{n-1}) \ldots$ and $\lim_{n \to \infty} V(c^n) = S$. The $\{c^n\}$ have a weak limit in F: Let this be $\bar{c}$. By the Banach-Saks Theorem [14] there exists a subsequence $\{c^m\}$ of $\{c^n\}$ such that

$$\{\frac{c^1 + c^2 + \ldots + c^m}{m}\} \xrightarrow{\ell_2} \bar{c}$$

i.e., whose means converge strongly to $\bar{c}$. Now by convexity

$$\{\frac{c^1 + c^2 + \ldots + c^m}{m}\} \in F$$

and by concavity of V,

$$S \geq V(\frac{c^1 + c^2 + \ldots + c^m}{m}) \geq \frac{V(c^1) + V(c^2) + \ldots + V(c^m)}{m}.$$

By the Banach-Saks Theorem on the line

$$\lim_{m \to \infty} \frac{V(c^1) + V(c^2) + \ldots + V(c^m)}{m} = S.$$

Hence

$$\lim_{m \to \infty} V(\frac{c^1 + c^2 + \ldots + c^m}{m}) = S ,$$

and the supremum S of V over F is attained at a point in F. ∎

Theorem 1 makes it clear that difficulties arise in the infinte-horizon case considered earlier because we did not put enough conditions on $B_t(c_t)$ to ensure that $V = \sum_{t=1}^{\infty} B_t(c_t)$ was continuous in the ℓ_2 norm. Our next step is therefore to characterize the suitable B_t's:

THEOREM 2 (Chichilnisky). *The real-valued function* $V(c) = \sum_{t=1}^{\infty} B_t(c_t)$ *is* ℓ_2-*continuous iff*

$$|B_t(c_t)| \leq b_t + \alpha c_t^2$$

where α *is a positive number and* $b \in \ell_1^+$, *the nonnegative cone of the space* ℓ_1 *of summable sequences*.

PROOF: See [3, Proposition 1] or [5, Proposition 2.1]. ∎

Theorem 2 requires the function $B_t(c_t)$ to be majorized and minorized by quadratics in c_t with positive and negative intercepts respectively, as in Figure 1. These intercepts are $\pm b_t$: as $b \in \ell_1^+$, $\lim_{t \to \infty} b_t = 0$ and the vertical intercepts of the majorizing and minorizing quadratics tend to zero.

REMARK 1. *Let* $B_t(c_t)$ *be concave*, c^1 *and nondecreasing and satisfy the conditions of Theorem 2. Then* $\lim_{t \to \infty} B_t(c_t) \leq 0, = 0$ *for all* $c_t \geq 0$.

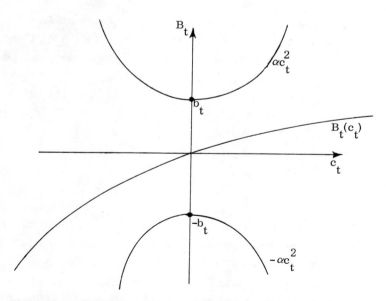

Figure 1: A function $B_t(c_t)$ satisfying Theorem 2.

PROOF: By Theorem 2, as $b \in \ell_1^+$,
$$\lim_{t \to \infty} |B_t(c_t)| < \alpha c_t^2$$

or
$$-\alpha c_t^2 \leq B_t(c_t) \leq \alpha c_t^2 .$$

Hence $B_t(0) \to 0$, and differentiating, in the limit we have
$$0 \leq \frac{dB_t(0)}{dc_t} \leq 0 .$$

Concavity implies that $d^2 B_t/dc_t^2 \leq 0$ so that dB_t/dc_t is nonincreasing. But $B_t(c_t)$ is nondecreasing means that $dB_t/dc_t \geq 0$. ∎

Theorem 2 thus implies that if $V(c)$ is to be ℓ_2-continuous, then the period-by-period payoff functions must eventually become flat: future consumption must give negligible benefits. Notice that $B_t(c_t) = B(c_t)$ concave and increasing, does not satisfy the conditions of Theorem 2. However, $B_t(c_t) = B(c_t)\delta_t$, $0 < \delta < 1$, may do so. Effectively, Theorem 2 implies that to obtain a solution to our optimal depletion problem, we have to discount future benefits relative to present ones.

Two general observations need to be made about Theorems 1 and 2.

REMARK 2. One would normally seek to show the existence of a maximum by demonstrating that the problem takes the form of maximizing a continuous function on a compact set. In this case it seems impossible to establish compactness and continuity in the same topology, compactness being elusive in infinite-dimensional spaces. The proof therefore shows that a continuous concave function attains its maximum on a closed bounded convex set: convexity is used heavily.

REMARK 3: It might seem natural to look at F as a subset of ℓ_1, the space of summable series. However, continuity of V in the ℓ_1-norm is not sufficient to ensure existence. To see this, consider the function

$$V(c) = \sum_{t=1}^{\infty} \alpha_t c_t$$

where $0 \leq \alpha_t \leq 1$, $\ldots \alpha_t < \alpha_{t+1} < \alpha_{t+2} \ldots$. This is ℓ_1-continuous, concave and increasing, but assumes no maximum on F, even though F is ℓ_1-closed, ℓ_1-bounded, and convex. Existence requires more than ℓ_1-continuity: also the existence proof requires us to work in ℓ_p, $1 < p < \infty$, rather than ℓ_1, as the Banach-Saks Theorem is only available for reflexive spaces.

REMARK 4. Aumann and Perles [1] and Artstein [2] have considered a related mathematical problem arising in economics.

The conclusion established -- that if the problem is well-posed, then we must give less weight to future than to present benefits -- is one that some commentators have found unappealing. Noteworthy in this respect is Ramsey's [13, p. 543] comment that "discounting of future utilities is ethically indefensible,...and arises merely from a weakness of the imagination." Harrod described discounting as representing "the victory of rapacity over reason." However, we have seen that if the problem is to be well-posed, then there is no choice but to discount. The only alternative is to look for a completely different way of generating an ordering over depletion paths. Some different approaches are suggested in [16] and reviewed in [8]. Solow [16] suggests as a criterion the selection of that depletion policy giving the highest sustainable level of consumption. In the present model the only path giving sustainable consumption is the trivial path $c_t = 0$ for all t, but in more complex models with alternative factors of production, this approach becomes more interesting. The basic idea is that although we may leave our successors little in the way of resources, we may be able to compensate by leaving them richly endowed with capital equipment and technological knowledge.

Before leaving altogether the question of the necessity of discounting, I want to make an important distinction, one which in some measure mitigates the more distasteful aspects of this conclusion. This distinction is between what are called the utility rate of discount and the consumption rate of discount.

The former is just the discount rate with which we are familiar already, and whose role was discussed above. It is the rate by which the weight attached to benefits declines with their futurity: it is called the utility rate of discount because economists often refer to the benefit function $B_t(c_t)$ as a utility function.

The consumption rate of discount is something that we have not yet encountered. It answers the following question. By what factor is the value of a small increment of consumption at date $t+1$ less than the value of the same increment at date t?
Formally, let

$$V(c) = \sum_{t=1}^{\infty} B(c_t)\delta^t$$

where $0 < \delta \leq 1$ and the function $B(c_t)$ is independent of t. Then we are interested in the ratio $(\partial V/\partial c_{t+1})/(\partial V/\partial c_t)$, which gives us the relative sensitivity of the maximand to increments of consumption at dates t and $t+1$. If this ratio is less than unity, then an increment of consumption is valued more highly if it is made available at an earlier date -- this is referred to as consumption discounting, and the ratio is then called the consumption discount rate.

Taking a second order approximation to $B(c_t)$ around c_t, we have

$$\frac{\partial V/\partial c_{t+1}}{\partial V/\partial c_t} = \delta(1 + \frac{\partial^2 B/\partial c_t^2}{\partial B/\partial c_t} \cdot (c_{t+1} - c_t))$$

as the consumption discount rate. Note that for linear benefit functions B, this is just δ: the two discount rates are the same. For the more general case of strictly concave benefit functions, the term in brackets exceeds one, and the consumption discount rate is less than the utility discount rate, whenever $c_{t+1} < c_t$. So on a path with declining consumption levels, the consumption discount rate is less than the utility discount rate. Clearly for a function B which is tightly curved, and on a path with rapidly falling consumption, one could have

$$\frac{\partial V/\partial c_{t+1}}{\partial V/\partial c_t} > 1:$$

future consumption would be given more rather than less weight than present, even though future benefits were discounted.

So the distinction between utility and consumption discount rates is important. The earlier mathematical arguments establish a case for utility discounting. This definitely need not imply consumption discounting, and in practical applications of project evaluation it is almost always consumption, rather than some more general measure of benefits, which is discounted. This is correct only if the implicit relation between benefits and consumption, and the sequence of consumption levels being studied, are such that

$$\frac{\partial V/\partial c_{t+1}}{\partial V/\partial c_t} < 1 \;.$$

This would be true for linear B functions: otherwise, there is work to do in establishing the validity of consumption discounting.

One final point deserves mention. Consider the problem (A) of selecting an optimal resource-depletion profile. Then it is easy to see that a necessary condition for a solution is that there exists a constant $\lambda > 0$ with $\partial B(c_t)/\partial c_t = \lambda$ for all t for which $c_t > 0$. In this case, the consumption discount rate is zero. So it is in fact a characteristic of an optimal profile that the consumption path be so chosen that the consumption discount rate is zero. The implications of this for the consumption path of course depend on the utility discount rate and the curvature of the benefit function.

3. PRICE-CHARACTERIZATION OF OPTIMAL DEPLETION PATHS

In many areas of economics, an important characteristic of an "optimal" policy is that it can be <u>decentralized</u> <u>via</u> <u>prices</u>. This means that there exists a set of prices such that if all agents act "rationally" at these prices -- sellers maximize profits and buyers maximize utilities subject to budget constraints -- then they act in such a way as to bring about the optimal policy. This is just an aspect of Adam Smith's "invisible hand" concept. I want now to show that this is also true of optimal depletion policies of the type discussed above.

Consider the problem "maximize $V(c)$ subject to $c \in F$," where F is as before and V is concave, ℓ_2-continuous and increasing. Increasing means that if $c_t^1 \geq c_t^2 \ \forall \ t$ and $\exists \ t' : c_{t'}^1 > c_{t'}^2$, then $V(c^1) > V(c^2)$. A <u>price</u> <u>vector</u> is a nonnegative sequence $(p_1, p_2, \ldots, p_t, \ldots)$ whose t-th component is the price of the exhaustible resource in period t. The <u>value</u> of a depletion sequence is then $\sum_{t=1}^{\infty} p_t c_t$: an obvious economic requirement is that this value be finite for any feasible depletion policy, i.e., any $c \in F$.

THEOREM 3. $c^* \in F$ <u>is an optimal depletion policy iff there exists a price vector</u> $p^* \in \ell_2$ <u>such that</u>

(i) $p^* \geq 0, \ p^* \neq 0$

(ii) $\langle p^*, c \rangle < \infty \ \forall \ c \in F$

(iii) $\langle p^*, c^* \rangle \geq \langle p^*, c \rangle \ \forall \ c \in F$

(iv) $\langle p^*, c \rangle \geq \langle p^*, c^* \rangle \ \forall \ c : V(c) > V(c^*)$

PROOF: (This follows that of Chichilnisky [5, Theorem 2]). Let $c^* \in F$ be optimal, and let $A = \{c \in \ell_2 : V(c) > V(c^*)\}$. By concavity and continuity of V, A is a convex set with nonempty interior. As V is increasing, $(\ell_2^+ + c^*) \subset A$. As c^* is optimal, $F \cap A = \emptyset$. Hence, by [10, Theorem 12, p. 412], there exists a nonzero linear functional p^* which separates F and A: as $(\ell_2^+ + c^*) \subset A$, this can be taken to be a positive linear functional. Hence by [15, Theorem 5.5, p. 228], p^* is continuous. This completes the proof of necessity. Sufficiency follows from convexity of F and concavity of V. ∎

REMARK 5. This theorem uses heavily the fact that the problem has been embedded in ℓ_2, which is a self-dual space. A linear functional on ℓ_2 is therefore representable by an inner product operation between two vectors in ℓ_2, and a linear functional has a natural economic interpretation as a price vector.

4. UNCERTAINTY

The choice of a depletion policy often seems, intuitively, to depend on issues about which we have little information -- will oil be discovered on the outer continental shelf? will there be a breakthrough in shale processing technologies? or in nuclear fusion? Here I want to review briefly how one can incorporate such issues into the problem studied above. (For rigorous treatments, see [7], [9] and [12].)

All of these possibilities suggest that at some unknown date T, the payoffs obtainable from using our resource will change dramatically -- in particular, they will drop, as substitutes will become available and its scarcity will diminish. It is natural, in picking the present depletion rate, to take this into account, but hard to know exactly how to do so, given the uncertainty about T.

One way of posing the problem is as follows. Let $\phi(t)$ be the probability that a major technological breakthrough will occur at date t, changing the payoffs to be obtained thereafter from our resource. Define:

$$W(S_T) = \text{maximum of } \sum_{T}^{\infty} B_t(c_t) \text{ if the breakthrough occurs at } T \text{ and stocks } S_T \text{ of the resource are then remaining.}$$

$W(S_T)$ is, in dynamic programming terms, a state valuation function. Now consider the problem

$$\text{"maximize } \sum_{T=1}^{\infty} \phi(T)\{\sum_{t=1}^{T} B_t(c_t) + W(S_T)\} \text{ subject to } c \in F." \tag{B}$$

This can be rewritten as

$$\text{"maximize } \{\sum_{t=1}^{\infty} B_t(c_t) \theta(t) + \sum_{t=1}^{\infty} \phi(t) W(S_T)\} \text{ subject to } c \in F." \tag{B'}$$

where $\theta(t) = \sum_{\tau=t}^{\infty} \phi(\tau)$. Note that as $\phi(t)$ is a probability density function, $\theta(t) \to 0$ as $t \to \infty$. It is therefore the case that $\sum_{t=1}^{\infty} B_t(c_t)\theta(t)$ may be ℓ_2-continuous even if $\sum_{t=1}^{\infty} B_t(c_t)$ is not. In problems (B) and (B'), the density function $\phi(t)$ is taken as exogenous. In fact, it can be influenced by present policies -- oil exploration, research and development, etc. Dasgupta, Heal and Pand (9) consider a refinement of this problem in which $\phi(t)$ can be influenced by current policies, and for a rather more complex model than that presented here obtain analytical results and present numerical simulations.

BIBLIOGRAPHY

1. R. J. Aumann and M. Perles, "A variational problem arising in economics", J. Math. Anal. Appl., 11 (1965), 488-503.
2. Z. Artstein, "On a variational problem", J. Math. Anal. Appl., 45 (1974), 404-414.
3. G. Chichilnisky, "Nonlinear functional analysis and optimal economic growth", J. Math. Anal. Appl., 61 (1977), 504-520.
4. G. Chichilnisky and G. M. Heal, "Existence of competitive equilibrium in Hilbert spaces", discussion paper no. 222, Columbia University, Department of Economics (1983), J. Econ. Theory., forthcoming.
5. G. Chichilnisky and P. J. Kalman, "Application of functional analysis to models of efficient allocation of economic resources", J. Optimization Theory and Applications, 30 (1980), 19-32.
6. G. Chichilnisky and P. J. Kalman, "Comparative statics and dynamics of optimal choice models in Hilbert spaces", J. Math. Anal. Appl., 70 (9179), 490-504.
7. P. S. Dasgupta and G. M. Heal, "Optimal depletion of exhaustible resources", Rev. Econ. Stud., 41 (1974), 3-28.
8. P. S. Dasgupta and G. M. Heal, Economic Theory and Exhaustible Resources, Cambridge University Press, Cambridge and New York (1979).
9. P. S. Dasgupta, G. M. Heal and A. Pand, "Funding research and development", Appl. Math. Modelling, (1980).
10. M. Dunford and J. Schwartz, Linear Operators, Interscience, New York (1958).
11. G. M. Heal, The Theory of Economic Planning, North Holland, Amsterdam (1973).
12. G. M. Heal, "Uncertainty and the optimal supply of an exhaustible resource", in Advances in the Economics of Energy and Resources, R. S. Pindyck, ed., J.A.I. Press (1977).
13. F. Ramsey, "A mathematical theory of saving", Econ. J., 38 (1928).
14. F. Riesz and B. Sz.-Nagy, Functional Analysis, Unger, New York (1955).
15. H. Schaefer, Topological Vector Spaces, Springer Verlag, Berlin and New York (1970).
16. R. M. Solow, "Intergenerational equity and exhaustible resources", Rev. Econ. Stud., 41 (1974), 29-40.

GRADUATE SCHOOL OF BUSINESS
COLUMBIA UNIVERSITY
NEW YORK, NY 10027

and

INSTITUTE FOR MATHEMATICS AND ITS APPLICATIONS
UNIVERSITY OF MINNESOTA
MINNEAPOLIS, MN 55455

Current Address:
Uris Hall
Columbia University
New York, NY 10027

Proceedings of Symposia in Applied Mathematics
Volume 32, 1985

CAPITAL THEORETIC ASPECTS OF RENEWABLE RESOURCE MANAGEMENT

Colin W. Clark[1]

ABSTRACT. A resource stock is a capital asset, differing only from capital as usually understood in economics by virtue of its internal dynamics. Any concept of optimal use over time leads to a dynamic optimization, or variational problem. This lecture describes a number of basic optimization models of renewable resources and their implications. Both deterministic and stochastic theories are described.

1. MYOPIC DECISION RULES. Consider the following dynamic optimization problem [2]:

$$\underset{\{h(t)\}}{\text{maximize}} \int_0^\infty e^{-\delta t} \pi(x,h) dt \qquad (1)$$

subject to

$$\frac{dx}{dt} = G(x) - h(t) \qquad (t \geq 0) \qquad (2)$$

$$x(0) = x_o \qquad (3)$$

$$0 \leq h(t) \leq h_{max} \qquad (4)$$

where

$$\pi(x,h) = p(x)h \qquad (5)$$

For the renewable resource setting, the symbols have the following interpretations: $x(t)$ is the resource stock size (biomass) at time t; $G(x)$ is the net rate of natural growth ($G(0) = G(K) = 0$; $G(x) > 0$ and $G''(x) < 0$ for $0 < x < K$); $h(t)$ is the harvest rate; $\pi(x,h)$ is the net revenue flow, or resource rent, which is proportional to the harvest rate but with a

---

1980 Mathematics Subject Classification: 90A16, 90B50, 92A15.
[1]Research supported by NSERC (Canada) under grant A-3990.

© 1985 American Mathematical Society
0160-7634/85 $1.00 + $.25 per page

nondecreasing coefficient $p(x)$, reflecting the increased costs of harvesting as the stock level x declines. The integral in (1) represents the present value of resource rents, discounted at the rate $\delta > 0$.

This singular variational problem can be solved by an elementary calculation. Write the integral as

$$\int_0^\infty e^{-\delta t} p(x) h \, dt = \int_0^\infty e^{-\delta t} p(x)(G(x) - \dot{x}) \, dt$$

$$= \int_0^\infty e^{-\delta t} W(x) \, dt + V(x_0) \tag{6}$$

where

$$W(x) = p(x)G(x) - \delta Z(x) \tag{7}$$

$$Z(x) = \int_{\bar{x}}^{x} p(y) \, dy \tag{8}$$

with $p(\bar{x}) = 0$, or $\bar{x} = 0$ if $p(x) > 0$ for all $x \geq 0$. Thus the problem is to determine the control $h(t)$, $t \geq 0$, that drives $x(t)$ so as to maximize the integral on the right side of (6).

If we assume that $W(x)$ is maximized over $[\bar{x}, K]$ at $x = x^*$, and that $W'(x) \gtreqless 0$ for $x \lesseqgtr x^*$, the solution becomes apparent: $x(t)$ should be driven as rapidly as possible towards x^*:

$$h^*(t) = \begin{cases} h_{max} & \text{if } x(t) > x^* \\ 0 & \text{if } x(t) < x^* \\ G(x^*) & \text{if } x(t) = x^* \end{cases} \tag{9}$$

Thus the optimal harvest policy is one that converges, as rapidly as possible, to an <u>optimal sustained yield</u> (OSY) $h^* = G(x^*)$, at a stock level characterized by the equation $W'(x^*) = 0$, or

$$G'(x^*) + \frac{p'(x^*)}{p(x^*)} G(x^*) = \delta \tag{10}$$

In the theory of capital and investment, Eq. (10) is known as the (modified) "Golden Rule of Capital Accumulation"; for $p' \equiv 0$ we obtain the ordinary Golden Rule

$$G'(x^*) = \delta \tag{11}$$

which asserts that the optimal size of a capital asset x is the level at which marginal productivity $G'(x)$ equals the discount rate δ. Sometimes the expression $G'(x)$, or more generally the expression on the left side of (10), is referred to as the own rate of interest of the asset x. In this form the rule was known to Keynes.

While the validity of the golden rule for optimal investment decisions is by no means universal, its scope can be extended considerably. Consider, for example, the case of time-dependent economic parameters $\delta = \delta(t)$, $p = p(x,t)$. It can be seen that (10) then becomes

$$G'(x^*) + \frac{p_x(x^*,t)}{p(x^*,t)} G(x^*) = \delta(t) - \frac{p_t(x^*,t)}{p(x^*,t)} \tag{12}$$

- the relative rate of price change is simply subtracted from the discount rate. Eq. (12) determines x^* as a function of t, referred to as the <u>myopic</u> solution (economics - [1]) or the <u>singular</u> solution (control theory). "Myopic" refers to the dependence of $x^*(t)$ only on current values of the parameters and their time derivatives - the future can be ignored.

But is the myopic solution optimal, even assuming (as we shall) that it is unique? Given $x^*(t)$, the corresponding harvest rate $h^*(t)$ is determined by our state dynamics (2):

$$h^*(t) = G(x^*(t)) - \frac{dx^*}{dt}$$

Obviously if $h^*(t)$ is not a feasible harvest rate (e.g., if $h^* < 0$) the myopic solution cannot be optimal. The converse also holds, in a sense [1]. Here is an example: assume δ = constant and

$$p(x,t) = \begin{cases} p_1 & 0 \leq t < T \\ \\ p_2 & t > T \end{cases} \tag{13}$$

where $p_1 > p_2$ (say), both being constants independent of x. The myopic solution $G'(x^*) = \delta$ holds for all $t \neq T$, but $x^*(t) \equiv x^*$ is not optimal (how should an asset owner behave when he knows that his asset is about to be devalued?).

The validity of a myopic investment rule depends on other assumptions also. For example, the function $W(x)$ must have a unique local maximum [13].

More fundamentally, the original optimization problem (1),(2) must be linear in the control variable. Finally, it would appear that the model must be a deterministic one; we discuss this later.

2. TWO ASSETS. Dynamic optimization problems in more than one state variable are nontrivial. We shall describe two resource models involving two state variables.

An obvious extension is to the case of two interacting species:

$$\frac{dx_i}{dt} = G_i(x_1, x_2) - h_i \qquad (i = 1, 2) \qquad (14)$$

We then have

$$\int_0^\infty e^{-\delta t} (p_1(x_1) h_1 + p_2(x_2) h_2) dt = \int_0^\infty e^{-\delta t} W(x_1, x_2) dt + \text{const} \qquad (15)$$

where

$$W(x_1, x_2) = \sum_1^2 [p_i(x_i) G_i(x_1, x_2) - \delta Z_i(x_i)] \qquad (16)$$

The solution appears to be "maximize $W(x_1, x_2)$ as rapidly as possible", but it is not clear what "as rapidly as possible" means. A conjectured solution involving a combination of bang-bang ($h_i = 0$ or max) and semisingular controls is stated in [2, p.323].

Suppose $W(x_1, x_2)$ is maximized at x_1^*, x_2^*, the OSY populations. The possibility arises that, e.g., $h_1^* = G_1(x_1^*, x_2^*) > 0$ even though $p_1(x_1^*) < 0$ --species x_1 may be a predator on the valuable species x_2. There are many real examples, often involving controversy between exploiters and conservationists. Some aspects of the multispecies fishery problem are described in [10].

Another two-variable model, rigorously solved in [4], is the following. Let $K = K(t)$ denote fishing capacity, e.g., the number of existing boats, and suppose in Eq. (4) that

$$h_{max} = qKx, \qquad q = \text{const.} \qquad (17)$$

Investment in boats can be increased by investment $I \geq 0$ (disinvestment is impossible); otherwise capacity depreciates at a constant proportional rate γ. Our model now becomes:

$$\underset{\{h(t),I(t)\}}{\text{maximize}} \int_0^\infty e^{-\delta t}[p(x)h - cI]dt \tag{18}$$

subject to

$$\frac{dx}{dt} = G(x) - h(t) \; , \qquad\qquad x(0) = x_o \tag{19}$$

$$0 \leq h(t) \leq qKx \tag{20}$$

$$\frac{dK}{dt} = I(t) - \gamma K \; , \qquad\qquad K(0) = K_o \tag{21}$$

$$0 \leq I(t) \tag{22}$$

The investment constraint (22) precludes the use of elementary methods to solve this optimization problem. Without the investment constraint, an integration by parts removing the variable I from (19) reduces this to the previous problem, but with $p(x)$ replaced by

$$p_1(x) = p(x) - \frac{c(\delta + \gamma)}{qx} \tag{23}$$

The second term simply represents interest and depreciation costs, which become variable harvest costs when I is completely reversible.

For the constrained problem it turns out that the optimal long-run equilibrium is again given by the solution x^* of Eq. (10), but with net revenue $p_1(x)$ of (23). For $x_o > x^*$, however, the optimal approach to equilibrium passes through a cycle of temporary overinvestment and overfishing. We refer to [4] for details, including economic interpretation and rigorous proof. The model is applied to the history of Antarctic whaling in [6].

3. COMMON-PROPERTY RESOURCES. The foregoing treatment of renewable resource economics assumes the existence of a resource owner, who is able to manage the resource to meet his financial objectives. Such property rights are common in some resource industries such as agriculture, but rare in others, such as wildlife and marine fisheries. The management of common-property resources to meet desirable social objectives is difficult, and often laced with controversy and multiple conflicts of interest [8]. If use of the resource is not regulated, basic economic principles predict overexploitation, with the stock reduced to a level at which net economic yield vanishes [7]. Effective methods of overcoming this "tragedy of the commons" in cases where property rights allocations are not feasible, are seldom obvious, and may not exist at all.

A mathematical analysis of common-property fishery regulation, based on models related to those presented here, appears in [3].

4. FLUCTUATIONS AND UNCERTAINTY. The annual recruitment of young fish to a given population often fluctuates by as much as one or two orders of magnitude. On the other hand, estimation of the current size of a fish stock in the ocean is obviously subject to considerable inaccuracy. Deterministic models of fisheries must therefore be treated with some suspicion.

The following simple discrete-time, stochastic fishery model has been studied by Reed [11]:

$$X_{k+1} = Z_{k-1} G(S_k) \qquad k = 1, 2, \ldots \qquad (24)$$

$$S_k = X_k - H_k \qquad (X_1 \text{ given}) \qquad (25)$$

where X_k denotes the stock level at the beginning of year k, H_k is annual catch, leaving spawning escapement S_k. The spawner-recruit relationship $G(S)$ is multiplied by a random factor Z, and we assume the Z_k to i.i.d. with common density ϕ. The optimization objective is the expected present value of economic rents:

$$\underset{\{H_k\}}{\text{maximize}} \ E\{\sum_1^\infty \alpha^k \pi(X_k, H_k)\} \qquad (26)$$

where

$$\pi(X_k, H_k) = \int_{X_k - H_k}^{X_k} p(x)\,dx \qquad (27)$$

and where $\alpha \in (0,1)$ denotes the discount factor.

The specific form of (27) renders the above model myopic. We have

$$E\{\sum_{k=1}^\infty \alpha^k \pi(X_k, H_k)\} = E\{\sum_{k=1}^\infty \alpha^k [Q(Z_k G(S_{k-1})) - Q(S_k)]\}$$

$$= Q(X_1) + E\{\sum_{k=1}^\infty \alpha^{k-1} [Q(Z_{k+1} G(S_k)) - \alpha Q(S_k)]\} \qquad (28)$$

where $Q(x)$ is an integral of $p(x)$. Define S^* by:

$$S^* \text{ maximizes } E\{Q(zG(S)) - \alpha Q(S)\} \qquad (29)$$

Then (28) leads to the guess that S^* is the optimal escapement level, and the optimal harvest policy is given by

$$H_k^* = \max(X_k - S^*, 0) \tag{30}$$

That this is indeed correct is established rigorously, under certain conditions, in [11] and [12]; the main requirement is that S^* is "self-sustaining" in the sense that

$$\Pr(zG(S^*) \geq S^*) = 1 \tag{31}$$

But even if S^* is not self-sustaining, the optimal harvest policy is still a constant-target-escapement policy of the form

$$H_k^* = \max(X_k - \bar{S}, 0) \tag{32}$$

for some $\bar{S} \geq S^*$ [11].

It may appear from the discussion so far that constant-escapement management policies are of general validity. Such policies are widely employed in fisheries management. In practice, economic questions are usually ignored; this amounts to setting $Q(S) \equiv S$ and $\alpha = 1$ in Eq. (29), which then reduces to the deterministic MSY (Maximum Sustained Yield) rule

$$G'(S) = 1 \tag{33}$$

Obviously any uncertainty about the average stock-recruitment relationship $G(S)$ will introduce corresponding uncertainty as to the optimal breeding stock S^*. But in the case of fish populations at least, <u>gross uncertainty is the primary characteristic of the relationship between stock and recruitment</u>. Except for certain extremely depleted stocks, the data often best supports a hypothesis of <u>no</u> relationship between stock and subsequent recruitment. Biologically there must of course be some connection, but any attempt to estimate parameters of a particular model invariably results in very wide confidence limits.

The question of optimal management policy with imprecise parameter estimates has been studied, using Bayesian methods, by Ludwig and Walters [9], who show that an "experimental" management approach may be required, in which escapement levels are deliberately varied over a wide range in order to improve parameter estimates. Such a policy is quite the opposite of a constant-escapement policy.

Another important source of uncertainty concerns the size of the resource stock at any given time. In the case of marine fisheries, confidence limits of at least ±50% in stock assessments are normal, and even optimistic for many species. Let us reconsider the model of this section, dropping the (tacit) assumption that the stock level X_k will be known exactly when the harvest quota H_k is determined as in Eq. (30). To avoid nonlinear filtering, assume that the previous season's escapement S_{k-1} is known exactly, but only at the end of the previous fishing season.

Under these assumptions X_k becomes a random variable with density $g_k^{-1} \phi(g_k x)$, where $g_k = G(S_{k-1})$. Following the method of dynamic programming, we define the value function $J_n(S_o)$ as the maximum expected present value, given previous escapement S_o, and with n periods remaining in the decision horizon. Assume now that

$$\pi(X, H) = H$$

We then have

$$J_1(S_o) = E\{X_1\} = G(S_o) \tag{34}$$

$$J_{n+1}(S_o) = \max_{Q \geq 0} E_{X_1} \{<X_1, Q> + \alpha J_n(X_1 - <X_1, Q>)\} \tag{35}$$

where $<x,y> = \min(x,y)$; this formulation assumes that the full quota Q will be harvested if $Q \leq X_1$, and otherwise the stock X_1 will be harvested entirely. The model captures the possibility of fishery collapse in the event of a misspecified quota, an important consideration for certain fisheries (such as the Peruvian anchoveta fishery).

For comparison, let $\tilde{J}_n(X_1)$ be the value function for the case of recruitment certainty, as in Reed's model discussed above. Then

$$\tilde{J}_1(X_1) = X_1 \tag{36}$$

$$\tilde{J}_{n+1}(X_1) = \max_{0 \leq H \leq X_1} [H + \alpha E_Z\{\tilde{J}_1(ZG(X_1 - H))\}] \tag{37}$$

Clearly $\tilde{J}_n(\bar{X}_1) \geq J_n(S_o)$ - the difference represents the value of complete recruitment information.

By means of the substitution $S = X_1 - Q$ we obtain

$$\tilde{J}_{n+1}(X_1) = X_1 + \max_{0 \leq S \leq X_1} [\alpha E_Z\{\tilde{J}_n(ZG(S))\} - S] \tag{38}$$

and a moment's consideration shows that the

optimal policy is a constant-target-escapement policy $S = S_n^*$, where S_n^* is independent of X_1 (but may depend on n). No such substitution is possible in Eq. (35), and it turns out in that case that the optimal policy is not a constant-escapement policy.

Numerical results for the two models are presented and compared in [5]. An unexpected outcome is that at high levels of stock fluctuation (about 80% coefficient of variation and above) the optimal target escapement $\bar{S} = E\{X_1 - Q^*\}$ is not an increasing function of expected recruitment $\bar{X}_1$. In extreme cases the optimal first-period quota becomes infinite, guaranteeing extinction of the resource. A combination of large fluctuations and high current stock levels produces a bias towards immediate harvest "while it's there." This remains true even for arbitrary time horizons and zero discounting, and is only exaggerated by short horizons and positive discounting. As a general principle, uncertainty and time discounting can have similar implications for resource management, but the connection is by no means simple or straightforward.

The problems of uncertainty in resource management are both important and difficult, particularly so in terms of the real-world complexity of biological and economic systems. Much more research will be needed before we understand how to optimize resource decisions under uncertainty.

BIBLIOGRAPHY

1. K.J. Arrow, "Optimal capital policy, the cost of capital, and myopic decision rules", Ann. Inst. Stat. Math., 16 (1964), 21-30.

2. C.W. Clark, <u>Mathematical bioeconomics: the optimal management of renewable resources</u>, Wiley-Interscience, New York, (1976).

3. C.W. Clark, "Towards a predictive model for the economic regulation of commercial fisheries", Can. J. Fish. Aquat. Sci., 37 (1980), 1111-1129.

4. C.W. Clark, F.H. Clarke, and G.R. Munro, "The optimal exploitation of renewable resources: problems of irreversible investment", Econometrica, 47 (1979), 25-49.

5. C.W. Clark and G.P. Kirkwood, "Optimal harvest policies for uncertain renewable resource stocks", preprint (1984).

6. C.W. Clark and R.H. Lamberson, "An economic history and analysis of pelagic whaling", Marine Policy, 6 (1982), 103-120.

7. H.S. Gordon, "The economic theory of a common-property resource: the fishery", J. Polit. Econ., 62 (1954), 124-142.

8. G. Hardin and J. Baden, <u>Managing the commons</u>, W.H. Freeman, San Francisco, (1977).

9. D.A. Ludwig and C.J. Walters, "Optimal harvesting with imprecise parameter estimates", Ecol. Modelling, 14 (1982), 273-292.

10. R.M. May, J.R. Beddington, C.W. Clark, S.J. Holt, and R.M. Laws, "Management of multispecies fisheries", Science, 205 (1979), 267-277.

11. W.J. Reed, "Optimal escapement levels in stochastic and deterministic harvesting models", J. Envir. Econ. Manag., 6 (1979), 350-363.

12. M.J. Sobel, "Stochastic fishery games with myopic equilibria", in L.J. Mirman and D.F. Spulber (eds.) <u>Essays in the economics of renewable resources</u>, North Holland, Amsterdam, (1982), 259-268.

13. M. Spence and D. Starrett, "Most rapid approach paths in accumulation problems", Intern. Econ. Rev., 16 (1975), 388-403.

DEPARTMENT OF MATHEMATICS
THE UNIVERSITY OF BRITISH COLUMBIA
#121 - 1984 MATHEMATICS ROAD
VANCOUVER, B.C., CANADA   V6T 1Y4

Proceedings of Symposia in Applied Mathematics
Volume 32, 1985

APPLYING ABSTRACT CONTROL THEORY TO CONCRETE MODELS: A CASEBOOK

Frank H. Clarke[1]

ABSTRACT. Through a series of examples drawn mainly from resource theory and economics, the main methods of optimal control theory and the issues arising in their application to specific models are presented.

There are three central themes that pervade optimization of all kinds, whether the problem involves algebraic or differential equations, or arises from physics or finance. The first of these is *existence*, the question of whether the problem under consideration admits a solution. The second is the topic of *necessary conditions*, by which we seek to delimit the possible solutions. These two steps may be likened to a criminal investigation, in which the existence of a crime is first ascertained and then clues are gathered in an attempt to identify the perpetrator. The third central theme, that of *sufficient conditions*, is analogous to providing incontrovertible evidence that a specific suspect is in fact the culprit.

An optimization problem with which we are all familiar, one which we label P_f, is that of minimizing a real-valued function f of a real variable x. In this setting, existence would follow readily from simple growth and lower semicontinuity assumptions on f. The classical necessary condition, which presupposes that f is sufficiently differentiable, is the assertion that if x is a point at which f attains its minimum, then $f'(x) = 0$ and $f''(x) \geq 0$. We call any point (suspect) x satisfying these conditions an *extremal*, a generic term for any point satisfying the necessary conditions appropriate to whatever problem is under discussion. To round off the theory for P_f, we have the result that a certain strengthening of these necessary conditions results in a set of conditions sufficient to guarantee that a given point x corresponds to at least a local minimum of f. The requirement is that x be an extremal *and* that $f''(x)$ be strictly positive.

1980 Mathematics Subject Classification 49A01, 49B01, 90A16, 90B50.

[1] The support of the Natural Sciences and Engineering Research Council of Canada is gratefully acknowledged.

© 1985 American Mathematical Society
0160-7634/85 $1.00 + $.25 per page

As simple as it is, P_f will serve to illustrate the two main approaches to solving optimization problems. The first, the *deductive method*, hinges upon the following logical chain of reasoning:

(i) a solution is known to exist (existence theorem)

(ii) the necessary conditions apply (study the extremals)

(iii) x is the best extremal (comparison or elimination).

This rigorously identifies and convicts a given point x on purely circumstantial evidence.

The *inductive method* takes some specific point x (which may have been identified by necessary conditions, but also by guessing, physical or economic intuition, trial and error, symmetry considerations, etc.) and proceeds to apply sufficient conditions directly to x to reach the conclusion that x is the solution. For P_f, this might entail showing that x is an extremal and verifying that $f''(x)$ is positive; this would secure a local minimum. Another type of sufficient condition would consist of showing that f can be factored as follows:

$$f(y) = [g(y) - g(x)]^2 + c.$$

If this were the case, it would follow that x is a global minimizer. The same conclusion is available if it is known that the function f is convex: in that case any critical point is a global solution to P_f.

In this article we shall be dealing with optimal control problems, which are considerably more complex than P_f, but which are also approachable deductively or inductively. It is because of this greater complexity that certain dangers in the use of the deductive method become more prominent.

Consider for example the initial step (i) in the reasoning. If it is missing, then even if we apply (correct) necessary conditions and find that there is a unique extremal x, it does not follow that x solves the problem, for there may not be a solution. (We would not wish to imprison someone on the basis of a confession to a nonexistent crime.) This consideration is usually not a source of difficulty for the simple problem P_f, in which existence is usually obvious. In contrast, the existence question for control problems is a subtle one, and reasonable-looking problems may not admit solutions. In practice, it is not uncommon to find the existence question "resolved" by appealing to the fact that the problem arises from a mathematical model of a physical or economic situation, and that the "real-world" problem certainly has a solution. This is a dangerous fallacy, for it is virtually equivalent to the statement "the model is a good one," which is what it is required to prove, not simply assert.

Another potential hazard in applying the deductive method lurks in step (ii) of the reasoning. Perhaps because of the complexity of control problems, or because the field is interdisciplinary, there exist in the literature incorrect necessary conditions ranging from the technically deficient to the totally absurd. Further, even correct necessary conditions only apply in the presence of the hypotheses under which they were derived. If for whatever reason only an uncertain characterization of extremals is available, then the deductive approach by itself cannot lead to a certain resolution of the problem. It is for reasons such as these that the inductive approach assumes greater importance in dynamic optimization than in the elementary problem P_f.

Just as there are types of necessary conditions in optimal control and the calculus of variations which find no counterpart in P_f, there exists a variety of existence theories and inductive methods. This is not the place to attempt a treatise explaining all these things, but we hope to communicate something of their nature and their role in the case histories to follow. To attain at least a measure of self-containment, we pause to state one version of the standard optimal control problem and the most commonly used tool in its analysis, the maximum principle of Pontryagin. (A full discussion of these and related topics appears in the author's book [8], whose notation is employed, and which contains all the mathematical theory we shall need.)

We consider the "controlled" initial-value problem

$$\dot{x}(t) = \phi(x(t),u(t)) \text{ a.e.}, \quad 0 \le t \le T, \quad x(0) = x_0,$$

where $\phi : R^n \times R^m \to R^n$ is a given function, T a given positive number and x_0 a given point in R^n. The function $u(\cdot)$, called the *control*, is one which we are free to choose, subject only to the conditions that it be piecewise continuous and assume values in a given subset U of R^m. Once $u(\cdot)$ is chosen, $x(\cdot)$ (the *state*) is defined as the solution to the differential equation. The object is to choose $u(\cdot)$ (and hence $x(\cdot)$) so as to maximize the functional $J(x,u)$ given by

$$f(x(T)) + \int_0^T F(x(t),u(t)) \, dt \, ,$$

where $f : R^n \to R$ and $F : R^n \times R^m \to R$ are two more given functions. We refer to this problem as P_C. For simplicity, we impose the following hypotheses: f is continuously differentiable, U is compact, ϕ and F are continuous and have derivatives in x which are continuous in (x,u).

The *maximum principle* is the necessary condition which asserts that if (x,u) is an optimal state/control pair for P_C, then there exists a function $p(\cdot)$ satisfying almost everywhere on $[0,T]$ the conditions

$$-\dot{p}(t) = D_x\phi(x(t),u(t))^* p(t) + D_x F(x(t),u(t))$$

(this is known as the *adjoint equation*; * denotes transpose) and

$$\max_{v \in U} \{p(t) \cdot \phi(x(t),v) + F(x(t),v)\} \text{ is attained for } v = u(t),$$

as well as the *transversality condition* $p(T) = Df(x(T))$.

As for existence, we can assert that P_C has a solution if we impose (for example) two further conditions: first, the *growth condition*

$$|F(x,u)| + |\phi(x,u)| \leq k(|x| + 1) \text{ for all } x, \text{ for all } u \in U;$$

and second the *convexity condition*: for each x, the following set is convex:

$$\{[\phi(x,u), F(x,u) - \delta] : u \in U, \delta \geq 0\}.$$

We now turn to case histories. These are all taken from the literature, and were motivated by "real world" problems, with the exception of the first, which is a technical exercise with a moral.

CASE HISTORY 1

The problem, a particular case of P_C, is to find the state/control pair (x,u) maximizing

$$J(x,u) := (x(1))^r/r + \int_0^1 e^{-x(t)^2/2} |u(t)| \, dt$$

subject to $\dot{x}(t) = u(t) \in [-1,1]$, $x(0) = 0$. Here r is an even integer satisfying

(1) $$\int_0^1 e^{-t^2/2} dt < 1 - 1/r$$

The conditions of the maximum principle reduce to

(2) $$\dot{p}(t) = x(t) e^{-x(t)^2/2} |u(t)|$$

(3) $$p(1) = x(1)^{r-1}$$

(4) $\max_{-1 \leq v \leq 1} \{p(t)v + e^{-x(t)^2/2}|v|\}$ is attained at $v = u(t)$.

Let us note that, without loss of generality, we may suppose that the optimal state is nonnegative on $[0,1]$ (this is a consequence of the symmetry of the problem: if $[a,b]$ is a subinterval on which x is nonpositive, with $x(a) = 0$, then replace (x,u) by $(-x,-u)$ on that subinterval without altering the objective functional). Further, we claim that the optimal state cannot be identically constant on a subinterval $[a,b]$. For if this is the case, let us redefine $u(\cdot)$ on $[a,b]$ to be (instead of zero) any nonconstant (continuous) function w with

$$\int_a^b w(t) \, dt = 0.$$

This alters x only on [a,b], and leads to a positive contribution from [a,b] in the objective functional integral, where before we had zero. Thus the new control would be an improvement, a contradiction.

We conclude from these observations, together with (2), that $\dot{p}(t)$ is positive a.e. The condition (4) implies that u(t) is +1 when p(t) is positive, -1 when p(t) is negative. We could not have p(0) negative, for then u(t) = -1 for t near 0, which would force x(t) to become negative. Thus $p(0) \geq 0$ and (since p is increasing) p(t) > 0 for t > 0. We deduce that u(t) is +1 for all t, so that x(t) = t. To confirm matters, we note (using (2)(3))

$$\dot{p}(t) = te^{-t^2/2}, \quad p(1) = 1,$$

which gives $p(t) = e^{-1/2} - e^{-t^2/2} + 1$, a function which is positive on [0,1] as we had derived.

The preceding derivation is not an unnatural or atypical analysis based on the maximum principle. It is not uncommon to see such a calculation followed by a conclusion, in this case that the unique state/control pair (t,1) identified by the reasoning is optimal; the resulting value of the objective J is then

$$1/r + \int_0^1 e^{-t^2/2} dt.$$

Consider however a function u(t) which switches back and forth between +1 and -1 so as to keep the value of the corresponding state x(t) very close to zero (of course, we could not have x identically zero this way). It is easy to see that by making this "chattering" sufficiently fast (and in consequence keeping x extremely near 0) we can make the value of J arbitrarily close to 1 (without being able to actually attain it). We now see from (1) that the control $u \equiv 1$ which we identified through the necessary conditions is decidedly not optimal.

The "only" fallacy in the argument was the (customarily tacit) assumption that a solution to the problem existed. The moral: beware of the deductive method (by itself) if existence is not assured.

Incidentally, the solution to this problem is a *relaxed control*, a mathematical idealization of a control which chatters infinitely fast; see [8, Section 5.5].

CASE HISTORY 2: OPTIMAL PRICING

Let x denote the (cumulative) quantity sold of a certain commodity by a firm facing a demand (rate) function $\phi(x,u)$, where u signifies the price of the commodity (u is chosen by the firm). This means that x grows according to the law

(1) $$\dot{x}(t) = \phi(x(t),u(t)), \quad 0 \leq t \leq T.$$

(It is natural to require $\phi_u < 0$.) The cost (rate) of producing q units of the commodity is specified by a function $c(x,q)$. The dependence of ϕ and c on x is present to enable the modeling of *experience effects* in demand and production, a general term encompassing phenomena such as consumer satiation, bandwagon effects, learning by doing, etc. Corresponding to a pricing policy $u(\cdot)$ on a time interval $[0,T]$ is a discounted net profit J given by

(2) $$\int_0^T e^{-\delta t}\{u(t)\phi(x(t),u(t)) - c(x(t),\phi(x(t),u(t)))\}dt,$$

where δ is the discount rate. Given an initial value $x(0) = x_0$, the problem is to choose the (piecewise continuous) pricing policy $u(\cdot)$ with values in $(0,\infty)$ which will generate a corresponding sales history $x(\cdot)$ via (1) so as to maximize the functional (2). (The problem, a central one in the theory of marketing, is discussed and analyzed in detail in [11], in which appears the first general analysis.)

The form of the problem is that of one in optimal control, although it differs from the simplified P_C of the introduction in that F has explicit dependence on t (i.e., the problem is nonautonomous) and the control region is not compact. Nonetheless under standard regularity hypotheses on the data the maximum principle does hold as stated [8, Theorem 5.2.1], and asserts that if (x,u) is a solution to this problem then there is a function p such that almost everywhere on $[0,T]$ one has

(3) $$\dot{p} = e^{-\delta t}\{c_x + c_q\phi_x - u\phi_x\} - p\phi_x$$

(4) $$\max_{0 < v < \infty} \{p\phi(x,v) + e^{-\delta t}[v\phi(x,v) - c(x,\phi(x,v))]\} \quad \text{at } v = u(t)$$

and such that $p(T) = 0$. The existence of an interior maximum in (4) implies

(5) $$p = e^{-\delta t}\{c_q - u - \phi/\phi_u\}.$$

It is not uncommon to see analyses of optimal control problems in which relations such as (5) and (3), which still involve the adjoint variable p, are the end product. When the goal is an "economic interpretation" of the answer, this is obtained by imputing to p an economic significance: p is a "shadow price" associated to the state x. What is accepted as constituting a solution of a given problem is conditioned by what is possible (a recurring theme in the case histories), and sometimes it may not be possible to take the analysis of the necessary conditions much beyond their statement. But this is rarely a satisfactory answer. In the present case, some finagling of (3) and (5) (see

[11]) eliminates the exogenous variable p to produce the following: along any optimal price path one has

$$(*) \qquad u(t) + \phi/\phi_u = c_q + \int_t^T (c_x + \phi\phi_x/\phi_u) e^{-\delta(\tau-t)} \, d\tau.$$

This turns out to be a relation fraught with economic significance. The left side of (*) is the quantity known as *marginal revenue* (MR) while the first term on the right is *marginal cost* (MC). The least that we all remember from our economics courses is the famous prescription for maximizing profit: equate MR and MC. We now see from (*) that when we are dealing with a dynamic problem in which an optimal policy is required over time (instead of a single myopic decision) and when the past affects the present through experience effects, then MR and MC will differ. The difference is the integral term in (*), which vanishes in the absence of experience effects (i.e., when $c_x = \phi_x = 0$). In [11], (*) is used to study special cases in which optimal price paths can be shown to be rising, falling or nonmonotonic.

In this example we have been using the deductive method, and we have obtained interesting information about solutions, without however having a complete description (for example, a formula). We have not discussed the existence question, an oversight which the lesson of the first case history will not allow us to commit, at least not in good conscience. It turns out that standard existence theorems do not apply to the problem, which considerably complicates matters. Nonetheless an ad hoc argument can be fashioned to prove the existence of solutions for a subclass of problems satisfying additional hypotheses; details are in [11, Appendix].

CASE HISTORY 3: OPTIMAL FISHING

In this model, the state x signifies the size of a certain biological population, a species of fish for example. The natural growth law of the population is

$$\dot{x}(t) = g(x(t)),$$

where g is a known function. We are able to apply effort towards harvesting the population: this is modeled by a function u(t) which we suppose constrained to a fixed interval [0,K]. The resulting growth (or diminution!) takes place according to the differential equation

$$(1) \qquad \dot{x}(t) = g(x(t)) - u(t)x(t).$$

The nature of the last term reflects the assumption that for a given effort rate the catch will be proportional to the number of fish present. The initial population size is a given positive quantity x_0. If we postulate a unit price π for fish and a unit cost c for effort, the net discounted revenue resulting from an effort rate u(·) applied over an interval [0,T] is given by

$$(2) \quad J(x,u) := \int_0^T e^{-\delta t}\{\pi x(t) - c\}u(t)\, dt,$$

where δ is again the discount rate. The problem, another instance of P_c, is to maximize this functional subject to the given constraints; a complete discussion and analysis are given by Clark in [5], see also [7]. As is frequently the case, we have here a tacit constraint imposed by the nature of the problem: $x \geq 0$. This kind of unilateral condition on the state is known as a *state constraint*. When state constraints exist in a control problem, the resulting version of the maximum principle is greatly complicated by the presence of measures in its statement [8, Theorem 5.2.1]. To avoid this, it would be helpful to show that the constraint is nonbinding, and therefore can be ignored for all intents and purposes. In our present situation, this means showing that any solution x remains strictly positive. This turns out to be an automatic consequence of the form of the differential equation (1), if we suppose that g(x) is nonnegative for x sufficiently small.

Standard existence theorems apply to this situation (see for example [8, Theorem 5.4.4]), so in applying the deductive approach we can now focus upon necessary conditions. There is a facet of this problem which is likely to simplify its analysis: both the right side of (1) and the integrand in (2) are linear in the control variable u. This property is inherited by the expression maximized in the maximum principle (the *pseudo-Hamiltonian*, see [8, Section 1.4]), which here is

$$pg(x) + v\{e^{-\delta t}(\pi x - c) - px\}.$$

The quantity σ in braces is called the *switching function*. Its significance is that u(t) must be 0 when σ is negative and K when σ is positive. If σ has only isolated zeros then u(·) must be a purely *bang-bang* control: one switching between the extreme points of the control set. In general however it will be possible for σ to vanish identically on a subinterval, which is then termed *singular*. The maximum principle then fails to immediately specify the value of u(·) on the subinterval. However, an analysis of the condition $\sigma = 0$, together with the state and adjoint equations, leads to the conclusion that on a singular subinterval the state value x must satisfy the algebraic equation

$$(3) \quad \frac{d}{dx}\{(\pi - c/x)g(x)\}/\delta = \pi - c/x.$$

As in the second case history, this relationship has an appealing economic interpretation in terms of marginal values. Under reasonable hypotheses on g, (3) has a unique solution $\tilde{x}$. The problem then reduces to deducing how best to combine bang-bang and singular subintervals.

An analysis of the conditions of the maximum principle [9] leads to the conclusion that the optimal policy is to control the population size as indicated in Figure 1 (we have chosen the case $x_0 > \tilde{x}$ for definiteness, and we

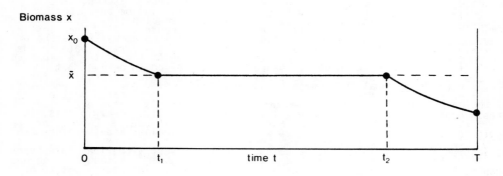

Figure 1

assume K sufficiently large). Thus we have $u = 0$ on $[0, t_1)$ and $u = K$ on $(t_2, T]$; in between, u is the intermediate value required to maintain $x = \tilde{x}$. The switching times t_1, t_2 are completely determined (although implicitly) by the two-point boundary-value problem for (x,p) resulting from the maximum principle (we know $x(0) = x_0$ and, by transversality, $p(T) = 0$). It is clear why in economics the type of conclusion we have reached regarding the optimal state is called a *turnpike theorem*.

In this example then, the deductive method has led to a complete solution of the problem.

Let us now consider the *infinite-horizon* problem in which $T = +\infty$, and see if the deductive method can be adapted to it. The existence of a solution becomes problematic, but assuming we can overcome that hurdle the next one would consist of finding a correct set of necessary conditions. It is natural to guess that the maximum principle holds essentially as before, with the transversality condition $p(T) = 0$ interpreted to mean $\lim_{t \to \infty} p(t) = 0$. Statements of this type exist in the literature, with the justification that since p is interpretable as a shadow price associated with the state, then (at least when the discount rate is positive, which we assume) it must tend to 0. A counterexample to this putative theorem is given in [3, Section 2], so the lesson may be drawn that obtaining mathematical results through economic interpretation is putting the cart before the horse.

The infinite-horizon problem admits a rigorous and elementary solution by the inductive method; this will be our first view of it in action. We first need a candidate for a solution. In view of the finite horizon analysis, a natural guess for the optimal state would be the one which goes to the value $\tilde{x}$ as fast as possible and then stays at $\tilde{x}$ forever. Let us call the control which accomplishes this $\hat{u}$, and the resulting state $\hat{x}$. Let (x,u) be any other admissible state/control pair. It turns out that because of the linearity in u, $J(\hat{x},\hat{u}) - J(x,u)$ can be viewed as a line integral over a certain path in the (t,x) plane, one to which Green's Theorem can be applied to derive that $J(\hat{x},\hat{u}) - J(x,u)$ is equal to a certain double integral over the region enclosed by the path. Under some hypotheses on g the integrand is everywhere nonnegative, which confirms that J attains a maximum at (x,u). Details of the argument, which is due to Clark, appear in [5].

Admittedly, there is an ad hoc flavor to the argument, although it does apply to a class of problems linear in the control, as shown in [5, Section 2.7]. This is not atypical of the inductive approach. Later examples will illustrate more systematic methods.

CASE HISTORY 4: RESOURCE EXTRACTION AND CAPITAL GROWTH

This model will feature a two-dimensional state $x = (y,z)$ and control $u = (v,w)$. The component y measures the current level of a certain (nonrenewable) resource (for example, gold), while z measures capital (machines) usable to extract the resource. Also involved is the labour v used in the extraction process and the labour w committed to building more machines. The dynamics are given by

(1) $$\dot{y} = -r(y) \min(z,v)$$

(2) $$\dot{z} = \gamma w,$$

where $r(\cdot)$ is a given function and γ a given constant. The factor $r(y)$ in the first equation reflects the fact that for a given extraction effort (i.e., z and v), extraction rates vary with the level of the remaining stock. The term $\min(z,v)$ reflects the situation wherein capital z and labour v must be used in fixed proportions ("one person, one shovel").

The total labour force $L(t)$ is exogeneously determined, hence the constraint

$$u = (v,w) \in U(t),$$

where $U(t)$ is the set determined by

$$0 \le v, \quad 0 \le w, \quad v + w \le L(t).$$

The central issue for the planner is to decide on a division of labour between that used for immediate production (v) and that used to build more capital stock (w). To put it another way, to what extent should present production be deferred in order to develop capital that will contribute to future production? The object is to maximize net discounted revenue, given by

$$(3) \qquad \int_0^T e^{-\delta t} \{\pi r(y) \min(z,v) - \alpha v - \beta w\} \, dt,$$

where T is the planning horizon, δ the discount rate, π the resource price, and α and β the labour costs in resource and capital production respectively.

This optimal control problem is one to which (under reasonable hypotheses) standard existence theorems can be applied [8, Proposition 3.2.3], so the next step in the deductive method is to find applicable necessary conditions. We turn to the maximum principle for help, but there is a problem. In the general terminology of P_C, the function $\phi(y,z,v,w)$ is given here by

$$[-r(y) \min(z,v), \gamma w],$$

a function which fails to be differentiable in the state variable (y,z). (For given v, the function $z \to \min(z,v)$ has a "corner" at $z = v$; the solution does in fact generally linger in this situation on a subinterval.)

In order to find an applicable theory, the problem can be rephrased in terms of *differential inclusions* (see [8, Chapter 3]). A differential inclusion is a relation

$$(4) \qquad \dot{x}(t) \in F(x(t)),$$

where F is a multifunction (i.e., set-valued function) from R^n to R^n. The central relationship between (4) and standard control problems with dynamics

$$(5) \qquad \dot{x} = \phi(x,u)$$

is that, under minimal hypotheses, if F is defined via $F(x) := \phi(x,U)$, then x satisfies (4) iff there is a measurable control u taking values in U for which (5) holds. (This is known as Filippov's Lemma.) Thus differential inclusions subsume classical control dynamics, and also serve as a suitable framework for more general situations. An important example is the case in which the control region U depends on x: U(x). This may be recast as a differential inclusion in the same way as above; we need only set $F(x) := \phi(x,U(x))$. It is decidedly *not* the case however that the maximum principle holds in this situation, despite assertions to this effect [2, Proposition 2].

The analysis of an optimal control problem framed via a differential inclusion (4) is carried out through the *Hamiltonian*

$$H(x,p) := \max\{<p,v> : v \in F(x)\}.$$

The crux of the necessary conditions is the assertion that if x is a solution, then a function p exists such that

(6) $$(-\dot{p}(t), \dot{x}(t)) \in \partial H(x(t), p(t)).$$

The symbol ∂ refers to the *generalized gradient* of H, a set-valued replacement for the usual gradient, necessitated under the present circumstances by the fact that H is in general nondifferentiable.

Space does not permit an extended discussion of generalized gradients or differential inclusions, for which we refer to [8]. Suffice it to say that our present problem can be cast as a differential inclusion problem to which the results of [8, Chapter 3] apply. These lead to a *Hamiltonian inclusion* (6) which can be shown to identify a unique strategy, which is therefore optimal (a success for the deductive method!). The model and its analysis are due to Clarke and Darrough [10]. It turns out that the optimal policy consists in general of three distinct phases (the first or third of which may be absent): an initial period of maximal production with all the remaining labour dedicated to building capital ($v = z(t)$, $w = L(t) - z(t)$), an intermediate period with no capital investment ($v = z(t)$, $w = 0$), and a period of shutdown ($v = w = 0$). The process is never fully capitalized: $z(t) < L(t)$, provided $z(0) < L(0)$. The switching times are fully (but implicitly) defined by a standard two-point boundary-value problem. We refer to [8, Section 3.3] for details and further remarks.

CASE HISTORY 5: REGULATING A DIODE

A well-known problem in optimal control theory that has found frequent application in engineering applications is the *linear regulator*, in which the goal is to achieve at minimum cost the transformation of the state x from an initial value x_0 at $t = 0$ to a target value x_T at $t = T$. The underlying dynamics are of the form $\dot{x} = Ax + Bu$, and the cost (of the control) is measured by a functional

$$J := \int_0^T \langle Qu(t), u(t) \rangle \, dt,$$

where Q is a positive definite matrix. The existence issue can be treated and the deductive method applied (via the maximum principle) quite satisfactorily to this problem.

Suppose now that the dynamics are provided by a simple circuit consisting of a capacitor, a diode and an impressed voltage in series (see [8, Figure 1.3, p.5]). Letting x signify the voltage across the capacitor and u the impressed voltage, we obtain the relationship

$$\dot{x} = \begin{cases} \alpha(u-x) & \text{if } x \leq u \\ -\beta(x-u) & \text{if } x \geq u \end{cases}$$

where α and β are positive constants. Supposing that $\alpha > \beta$, we may write this as follows:

(1) $$\dot{x} = \max\{\alpha(u-x), \beta(u-x)\}.$$

In the spirit of the linear regulator problem (but foregoing linearity), we consider the problem of minimizing $\int_0^T u^2 dt$ over those controls u (with values in R) whose associated state x attains prescribed values x_0, x_T at $t = 0, T$ respectively. In view of the right side of (1), we are once more in the situation in which the classical maximum principle is not applicable due to nonsmoothness of the data. A further complication is that the standard existence theorems do *not* apply to the problem, as the convexity condition fails.

We shall proceed with an eye to using the inductive method. The first step is to formulate a conjecture regarding the optimal control. To this end, we shall reformulate the problem as a generalized *problem of Bolza* (in the calculus of variations) and apply the necessary conditions developed in [8, Chapter 4].

We define functions L and ℓ by

$$L(x,v) := (x + v/\alpha)^2/2 \quad \text{if} \quad v \geq 0$$
$$:= (x + v/\beta)^2/2 \quad \text{if} \quad v \leq 0,$$

$$\ell(s_0, s_1) := 0 \quad \text{if} \quad s_0 = x_0, \quad s_1 = x_T$$
$$:= +\infty \quad \text{otherwise.}$$

An exercise for the reader: by solving for u in (1), show that the control problem is equivalent to that of minimizing

$$\ell(x(0), x(T)) + \int_0^T L(x, \dot{x}) \, dt$$

The use of extended-valued functions such as the ℓ above is a useful bookkeeping device to keep implicit account of constraints. The theory in [8] (which allows L to be extended-valued also) stresses the use of the *Hamiltonian* H corresponding to L, the function defined by

$$H(x,p) := \sup\{<p,v> - L(x,v)\}.$$

Under certain hypotheses (present in this case), it is possible to assert that corresponding to a solution x is a function p such that the following Hamiltonian inclusion holds

$$(-\dot{p}, \dot{x}) \in \partial H(x,p) \quad \text{a.e.,} \quad 0 \leq t \leq T,$$

as well as the condition $H(x(t), p(t)) = $ constant. (As in the previous example,

∂H refers to the generalized gradient of H.) It is not difficult to calculate H (see [8, Section 5.3]); its pattern of level curves is indicated in Figure 2; the arrows indicate the direction of motion of $(x(t),p(t))$ along such level

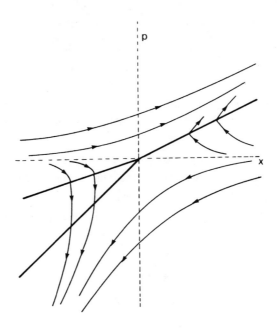

Figure 2

curves. The information so derived uniquely specifies a state $x(\cdot)$ for any pair (x_0, x_T) of endpoints. For example, suppose we have $x_T = x_0 < 0$. Then (x,p) must trace a path such as the ones in the third quadrant in Figure 2, in the course of which x in turn increases, remains constant, and then decreases. The behavior of x is completely determined by a two-point boundary-value problem; see [8, pp. 216-217] for details.

The necessary conditions have uniquely specified a strategy, so if the problem has a solution, that must be it. It remains now to confirm the optimality of that strategy. There is a recent set of sufficient conditions due to V. Zeidan [14] whose main hypothesis is that x be an extremal in a strengthened sense (i.e., that a stronger form of the Hamiltonian inclusion hold). It would take us too far afield to discuss this matter here, so we refer the reader to [8, Section 5.4] for a discussion of this topic. As shown there, the sufficient conditions apply to the problem at hand, providing an inductive finish to the analysis.

It is possible to extend the maximum principle to nondifferentiable data [8, Section 5.2] via generalized gradients. The resulting necessary conditions may be more or less precise than the Hamiltonian inclusion. In the present case, as shown in [8], the maximum principle fails to identify a single strategy. The moral to draw is that we are better off having several strings to our bow.

CASE HISTORY 6: RENEWABLE RESOURCES AND IRREVERSIBLE INVESTMENT

We consider now a model with two state components x and K representing respectively the population size of a renewable resource and the capital stock used in harvesting it, and two control components u and I representing respectively the harvesting effort and investment (rates). The dynamics are given by

$$\dot{x} = g(x) - ux$$
$$\dot{K} = I - \gamma K,$$

where γ is a positive parameter corresponding to the depreciation rate of capital. Note that the dynamics of the renewable resource are those of Case History 3, but that now capital is a state variable as well. The harvesting effort u is constrained to the (now state dependent) interval $[0,K]$, and we allow the investment I to assume values in $[0,\infty)$. In fact we allow $I = +\infty$, in a sense, by admitting the possibility of (upward) jumps in the value of K (this corresponds to an instantaneous purchase of additional capital).

The net discounted revenue corresponding to a policy $u(t)$, $I(t)$ of effort and investment and to purchases of capital stock of size ε_i at times t_i ($i = 1, 2, \ldots$) is given by

$$\int_0^\infty e^{-\delta t}\{(\pi x - c)u - rI\} dt - r\sum_i e^{-\delta t_i}\varepsilon_i,$$

where $r > 0$ represents the unit cost of investment.

The problem of course is to manage population size and capital so as to maximize the net revenue. It is a fact of life that the problem lies at the fringe of the class for which there is hope of finding an analytic solution. In particular, the task of using mere necessary conditions of maximum principle type to determine the answer more or less completely is hopeless. Perhaps the best indication of this is the answer itself, which is indicated in feedback form in Figure 3. In this figure, the central characters are certain values $\tilde{x}$, x^* of resource stock, a certain value K^* of capital stock, and two switching curves S_1, S_2. The curves bearing arrows are state trajectories.

To understand Figure 3, it suffices to consider the example in which the initial conditions $x(0)$, $K(0)$ situate us at the point marked A in the x-K plane.

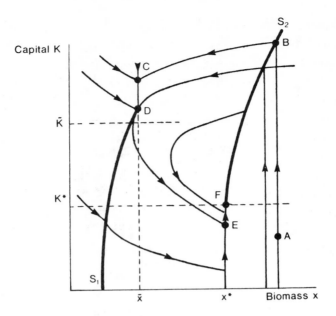

Figure 3

This is to be understood as a situation in which the population (fish) is relatively unexploited and little capital (boats) exists. The optimal strategy is to purchase boats at time $t = 0$ up to the level to bring us to the point B in the figure (i.e., up to S_2). Subsequently, we apply maximum harvesting effort ($u = K$) and allow boats to depreciate ($I = 0$) until the point C is reached. We then proceed to apply the precise level of effort needed to maintain the resource level at $\tilde{x}$, continuing to not invest, until we arrive at D. We then adopt the policy $u = K$, $I = 0$ until such time as $x = x^*$ is attained. (Boats have depreciated to the extent that the fish population cannot be held in check even with maximal effort; note however that x does dip below $\tilde{x}$ momentarily.) Once E has been reached, we purchase (instantaneously) the number of boats required to bring the state to the point F. From that time on, the continuous levels of effort and investment required to stay at F are applied.

We refer to [6] for further elaboration on the model. Suffice it to say for now that (x^*, K^*) represents the long-run optimal equilibrium when all is considered, whereas $\tilde{x}$ (which is calculated exactly as in Case History 3) represents a short-run optimal stock level when capital is not a factor (i.e., when boats are plentiful). For supplementary detail, and for an extension to the case of bounded investment, see [13].

How was the answer obtained? First the singular stationary values $\tilde{x}$ and x^* were found through analysis of the corresponding Hamiltonian inclusion. There are certain elements of the problem which make the application of that theory less than rigorous, but all is fair in formulating guesses. Next, economic intuition was used to guess the nature of the optimal strategy in a certain region bordering (x^*, K^*) (specifically, the region having $u = K$, $I = 0$). If the guess were correct, a certain necessary condition, a generalized Hamilton-Jacobi equation for a function generated by the guess, would be expected to hold. This condition fails along S_2 (which is how S_2 is defined), leading to a redefinition (a reguess?) of the optimal strategy (to $I = +\infty$) to the right of and below S_2. The process continues in this way until a global solution of the Hamilton-Jacobi equation is obtained, which in itself confirms the optimality of the guess. The approach is therefore seen to lie within the inductive method.

The Hamilton-Jacobi verification technique, within the context of the problem P_C described in the introduction, hinges upon the existence of a function $V(t,x)$ satisfying a condition of the type

$$\max_{v \in U} \{V_x(t,x) \cdot \phi(x,v) + F(x,v)\} + V_t(t,x) \leq 0.$$

We ask also that along the putative optimal pair (x,u) equality hold, that the maximum in question be attained at $v = u(t)$, and also that $V(T,\cdot)$ and $f(\cdot)$ coincide.

It is an easy matter to see that the existence of such a function V confirms the optimality of (x,u). For consider any other feasible state/control pair (y,v) for the problem. Then one has

$$V_x(t,y) \cdot \dot{y} + F(y,v) + V_t(t,y) \leq 0.$$

Integrating, the reader may confirm that we deduce

$$J(y,v) \leq V(0,x_0),$$

with equality if $(y,v) = (x,u)$. It follows that (x,u) maximizes J. The technique called *dynamic programming* is based on the same idea.

A reasonable question to ask is whether one can really expect to confirm optimality by such means. It turns out that when the method is extended to the nonsmooth setting through generalized gradients, the existence of a verification function V becomes a necessary and sufficient condition for optimality. We refer to [8, Section 3.7] for details, and to [1][6][12] for examples of applications.

We have not discussed any stochastic problems in this article, for it is usually not possible to find a complete analytic solution to them. This doesn't mean however that analytic solutions and the methods of this article cannot play a role in their analysis. An example of this is the version of the problem of Case History 6 incorporating stochastic elements treated by

Charles [4]. Practically speaking, the problem is amenable only to numerical solution, and the development of a good implementable algorithm is itself a formidable task. The analytic solution to the underlying deterministic problem is invaluable not only as a guide in designing the approximation process, but in providing assurance that the numerical scheme is producing good results (i.e., by shrinking the uncertainty and testing convergence to the known deterministic answer).

A final remark. Because of the structure of the problem (infinite horizon and merely exponential time dependence in the cost) it is possible to express the solution to the problem in *feedback* form (i.e., future policy depends only on the current state, and not on the current time). This is always the goal when it is possible; it would not be possible for the finite-horizon problem.

SUMMARY

There are not only ways to solve problems but ways not to solve them. The methods can be classified as deductive or inductive, and may lead to various types of conclusions which we may be willing to accept as solutions. The methods available, and what is required in the way of a solution are issues frequently linked to the modeling process itself. The solution of an interesting control problem is rarely routine, and is almost always endowed with an important ad hoc component. It is important to have the gamut of methods available, and these may necessitate nonstandard techniques (e.g. nonsmooth analysis) even when the underlying problem is quite naturally posed. Although we cannot always achieve one, nothing is quite as satisfying as a complete analytic solution.

REFERENCES

1. R.A. Adams and F.H. Clarke, Gross's logarithmic Sobolev inequality: a simple proof, Amer. J. Math. 101 (1979), 1265-1269.

2. K.J. Arrow, Applications of control theory to economic growth, in "Lectures in Applied Mathematics 12", Mathematics of Decision Sciences (G. Dantzig, A. Vienott, editors), Amer. Math. Soc., Providence, R.I., 1969.

3. J.P. Aubin and F.H. Clarke, Shadow prices and duality for a class of optimal control problems, SIAM J. Control Opt., 17 (1979), 567-587.

4. A.T. Charles, Optimal fisheries investment under uncertainty, Can. J. Fish. Aq. Science, 40 (1983), 2080-2091.

5. C.W. Clark, "Mathematical Bioeconomics", Wiley and Sons, New York, 1976

6. C.W. Clark, F.H. Clarke, and G.R. Munro, The optimal exploitation of renewable resource stocks, Econometrica, 47 (1979), 25-47.

7. C.W. Clark and G.R. Munro, Economics of fishing and modern capital theory: a simplified approach, J. Envir. Econ. Management, 2 (1975), 92-106.

8. F.H. Clarke, "Optimization and Nonsmooth Analysis", Wiley and Sons, New York, 1983.

9. F.H. Clarke, Lecture notes in the calculus of variations and optimal control, University of British Columbia (unpublished).

10. F.H. Clarke and M. Darrough, Resource extraction versus capital investment under fixed proportions: a nondifferentiable control problem, IAMS Technical Report No. 81-2, University of British Columbia, 1981 (excerpted in [8, Section 3.3]).

11. F.H. Clarke, M. Darrough and J.M. Heineke, Optimal pricing policy in the presence of experience effects, Journal of Business 55 (1982), 517-530.

12. F.H. Clarke and R.B. Vinter, On the conditions under which the Euler equation or the maximum principle hold, J. Appl. Math. Opt. (to appear).

13. G. Frenisy, Gestion optimale de pêcheries, thesis, Université de Paris IX, 1980.

14. V. Zeidan, Sufficient conditions for the generalized problem of Bolza, Trans. Amer. Math. Soc., 275 (1983), 561-586.

CENTRE DE RECHERCHE DE MATHEMATIQUES APPLIQUEES
UNIVERSITE DE MONTREAL
C.P. 6128, SUCCURSALE A
MONTREAL, P.Q., CANADA H3C 3J7

INTERNATIONAL TRADE IN RESOURCES:
A GENERAL EQUILIBRIUM ANALYSIS

G. Chichilnisky

Columbia University

and

Institute for Mathematics and its Applications
University of Minnesota

1. Introduction

Physical considerations alone cannot explain the volatile behavior of resource prices, or the effects these have on different regions of the world. An optimization analysis may not suffice either, since typically there are several distinct objectives: conservation, cost-minimization, and the maximization of revenues by resource exporters. These issues require an economic analysis of markets.

Markets for resources interact rather strongly with other markets: for goods such as food or industrial products and for inputs to production such as labor and capital. Such interactions are best studied with general equilibrium tools. These tools explain trade and the determination of prices across different markets. In a general equilibrium model, different economic agents have typically different objectives, a useful feature for the study of resource markets. Trade in resources takes place largely across different regions, so one is dealing with <u>international trade</u>.

A general equilibrium model of a market economy consists of a set of simultaneous, generally nonlinear, equations. A solution determines market prices and the quantities employed, produced, consumed and traded by the various agents in different markets. One aims to understand the qualitative changes of a solution consequent upon changes in the exogenous parameters of the system. This is generally accomplished by studying relationships between the variables within the manifold of solutions described as the exogenous parameter changes. This method is called "comparative statics": it seeks to explain what determines the outcomes, and how policies can be designed so as to obtain desirable solutions. The name "comparative statics" is somewhat misleading, since the same method is used to study the asymptotic behavior of dynamical systems in economics.

A system of simultaneous nonlinear equations can easily become unmanageable, and require computer analysis. Computer solutions cannot, however, disclose laws of economic behavior, nor can they explain why and how certain policies work. The challenge is therefore to represent the economy by a set of equations which is sufficiently simple to admit analytic or simple implicit solutions and the study of their qualitative behavior, while at the same time retaining the complexity needed to explore the issues involved. This paper will show how to perform this task and apply the results to study policy issues in the area of natural resources. We shall analyze:

1. The connections between resource exports, their prices, and the distribution of income within the exporting economy;

2. When a country should export more, and when less;

3. How the prices of resources affect employment, output and industrial prices within a resource importing region;

4. The effects of monopolistic resource prices on the prices of industrial goods which are traded in exchange for the resource:

5. Whether or not higher oil prices lead to lower GDP in an importing region, and whether there are cases where both the buyer and the seller of a resource may benefit from a given price change;

6. The effects of loans from a resource importer to a resource exporter when these are used to expand resource-extraction in the exporting region;

7. The effects of (unpaid) loans on the price of the resources exported;

8. The welfare effects of loans on the lending region: can an (unpaid) loan make the lender better off and the borrower worse off and if so, under what circumstances?

The following sections will present general equilibrium models of increasing complexity designed to analyze these questions, and theorems obtained on these issues.

2. The North-South Model

A. Specification of the Model

This section summarizes the general equilibrium model introduced in Chichilnisky (1981, 1983). There are two regions, North and South. The North represents the industrial countries, the South the developing countries. Each region produces and consumes two goods: basics (B) and industrial goods (I). There are two inputs to production: capital (K) and labor (L). Basics can also include a natural resource or a raw material. The two regions trade with each other.

Consider first the economy of the South. It produces basics and industrial goods using labor and capital, as described by the Leontief production functions

$$B^S = \min(L^B/a_1, K^B/c_1)$$

$$I^S = \min(L^I/a_2, K^I/c_2) \, ,$$

where the superscripts B and I denote the sector in which inputs are used, and the superscript S denotes supply. Efficiency requires that firms always use factors in fixed proportions:

$$L^B/K^B = a_1/c_1 \quad \text{and} \quad L^I/K^I = a_2/c_2 \, .$$

We can now write the equations that specify the model. Competitive behavior on the part of the firms ensures zero profits, so that

$$p_B = a_1 w + c_1 r \qquad (2.1)$$

$$p_I = a_2 w + c_2 r \qquad (2.2)$$

where p_B and p_I are the prices of B and I; w and r are the wages and the rate of return on capital. Equations (2.1) and (2.2) embody the information given by the production functions for B and I.[1]

---

[1] Equations (2.1) and (2.2) are equivalent to the production functions when firms

Labor and capital supplied are increasing functions of their rewards:

$$L^S = \alpha(w/p_B) + \bar{L} \qquad (\alpha > 0), \qquad (2.3)$$

$$K^S = \beta r + \bar{K} \qquad (\beta > 0). \qquad (2.4)$$

We now give market clearing conditions (superscript S denotes supply and D denotes demand):

$$L^S = L^D \qquad (2.5)$$

$$K^S = K^D \qquad (2.6)$$

$$L^D = L^B + L^I = B^S a_1 + I^S a_2 \qquad (2.7)$$

$$K^D = K^B + K^I = B^S c_1 + I^S c_2 \qquad (2.8)$$

$$B^S = B^D + X_B^S, \text{ where } X_B^S \text{ denotes exports of B} \qquad (2.9)$$

$$I^D = X_I^D + I^S, \text{ where } X_I^D \text{ denotes imports of I,} \qquad (2.10)$$

and

$$p_B X_B^S = p_I X_I^D, \qquad (2.11)$$

i.e., the value of exports equals the value of imports.[2]

---

are competitive. Profits are zero in this case, which means revenues equal costs, i.e., $p_B B^S = wL^B + rK^B$. Now from the production functions $B^S = L^B/a_1 = K^B/c_1$, so that $p_B B^S = a_1 w B^S + c_1 r B^S$, or $p_B = a_1 w + c_1 r$, equation (2.1). Similarly, one derives equation (2.2).

[2] It is worth noting that in this model when all markets clear, the value of domestic demand $p_B B^D + p_I I^D$ equals the value of domestic income $wL + rK$. This is called Walras' Law or the national income identity. From (2.1)-(2.11) one obtains:

$$p_B B^D + p_I I^D = p_B(B^S - X_B^S) + p_I(I^S + X_I^D) = p_B B^S + p_I I^S = (a_1 w + c_1 r) B^S + (a_2 w + c_2 r) I^S$$

$= wL + rK$. In view of this, and its homogeneity properties (solutions only depend on relative prices) the model is consistent with a standard Arrow-Debreu general equilibrium model, for some set of preferences.

The North is specified by a similar set of equations (2.1) - (2.11), with possibly different technology and factor supply parameters. In a world equilibrium, the prices of traded goods are equal across regions (factors K and L are not traded) and exports match imports:

$$p_I(S) = p_I(N) \tag{2.12}$$

$$p_B(S) = p_B(N) \tag{2.13}$$

$$X_B^S(S) = X_B^D(N) \tag{2.14}$$

$$X_I^S(N) = X_I^D(S) , \tag{2.15}$$

where letters S and N in brackets denote South and North respectively.

In each region there are eight exogenous parameters: a_1, a_2, c_1, c_2, α, $\bar{L}$, β and $\bar{K}$, making a total of <u>sixteen exogenous parameters</u> for the North-South model. When we add the price normalizing condition,[3]

$$p_I = 1 \tag{2.16}$$

we have a <u>total of 26 independent equations</u>: (2.1) - (2.11) for North; (2.1) - (2.11) for South, (2.12) through (2.14) and (2.16).[4] There are in total <u>28 endogenous variables</u>, 14 for each region: w, r, L^S, L^D, K^S, K^D, B^S, B^D, X_B^S, I^S, I^D, X_I^D, p_B, p_I. Therefore the system is underdetermined up to two variables.[5] Thus, we now specify two more variables exogenously, industrial demand in the South, $I^D(S)$ and in the North, $I^D(N)$. Obviously we could have solved the model by specifying other variables, or else by postulating demand equations; this will be done in the following sections. The demand specifications of the model are chosen to meet two criteria: analytical tractability and empirical plausibility.

---

[3] All relations in this model are homogeneous of degree zero in prices, i.e., only relative prices (rather than the price level) matter. Therefore, one normalizes the model by fixing the price of one good (the numeraire) equal to 1.

[4] It is easy to see that (2.15) is always satisfied when (2.11) is satisfied in each region and (2.12) through (2.14) hold.

[5] This is not surprising, since demand behavior, or preferences, have not been specified so far.

The North-South model is, therefore, a system of 26 equations in 26 variables, depending on 18 exogenous parameters: a_1, a_2, c_1, c_2, α, $\bar{L}$, β, $\bar{K}$ for each region, plus $I^D(S)$ and $I^D(N)$.

B. Comparative Statics Results

The economies of the North and of the South are identical except possibly for the values of their exogenous parameters. Differences in the structural characteristics of the two regions are described by differences in their exogenous parameters. For instance, in the North the two sectors (B and I) use approximately the same technology, i.e., the economy is technologically homogeneous. This means that $a_1/c_1 \sim a_2/c_2$ so that the determinant $D(N)$ of the matrix of technical coefficients

$$\begin{pmatrix} a_1 & c_1 \\ a_2 & c_2 \end{pmatrix}$$

is close to zero in the North. In the South, instead, technologies are dualistic. The two sectors use factors very differently, and $D(S)$ is therefore large. In both regions $D(N)$ and $D(S)$ are positive, which indicates that the B-sector uses labor more intensively than the I-sector. Another difference arises in factor markets. In the North labor is relatively more scarce, i.e., less responsive to increases in the real wage w/p_B. This means $\alpha(N)$ is small. In the South the opposite is true, $\alpha(S)$ is large. The reciprocal relations hold in capital markets: $\beta(N)$ is large, and $\beta(S)$ is small. These parameter specifications can be presented so as to be independent of the units of measurements.

It is worth noting that while most equations are linear in the variables, some are not (e.g., (2.3) is nonlinear). The solutions also display nonlinearities, as we shall see in the following.

Recall that a <u>North-South economy</u> is defined by: equations (2.1) through (2.11) for each region, and (2.12) through (2.15). Exogenous parameters are a_1, c_1, a_2, c_2, α, β, $\bar{L}$, $\bar{K}$ for each region, and $I^D(S)$, $I^D(N)$. Endogenous parameters are p_B, p_I, w, r, L^S, L^D, K^S, K^D, B^S, B^D, X_B^S, I^S, I^D, X_I^D for each region.

Therefore there are twenty-six independent equations in twenty-six variables, and eighteen exogenous parameters.

Proposition 1. <u>A North-South economy has at most one solution. This solution can be computed explicitly by solving one equation which depends on all exogenous parameters of the model.</u>

<u>Proof.</u> From

$$X_I^D(S) = X_I^S(N)$$

we have

$$I^D(S) - I^S(S) = I^S(N) - I^D(N) \qquad (2.17)$$

Inverting (2.7) and (2.8) we obtain

$$B^S = \frac{c_2 L - a_2 K}{D} \qquad (2.18)$$

$$I^S = \frac{a_1 K - c_1 L}{D} \qquad (2.19)$$

and inverting (2.1) and (2.2),

$$w = \frac{p_B c_2 - c_1}{D} \qquad (2.20)$$

$$r = \frac{a_1 - p_B c_2}{D} . \qquad (2.21)$$

We now rewrite (2.17) as a function of one variable only, p_B (which is the "terms of trade" of the South, since $p_I = 1$) and obtain:

$$p_B^2(A + A(N)) + p_B[C + C(N) + I^D(S) + I^D(N)] - (V + V(N)) = 0 \qquad (2.22)$$

for

$$A = \beta a_1 a_2 / D^2$$

$$V = \alpha c_1^2 / D^2$$

and

$$C = \frac{1}{D}\left(c_1 \bar{L} - a_1 \bar{K} + \frac{(\alpha c_1 c_2 - \beta a_1 a_2)}{D}\right) ,$$

and where A, V and C have parameters for the South, and A(N), V(N) and C(N) for the North. Solving equation (2.22) yields p_B^* as a function of all exogenous parameters of the system.

It is easy to check that (2.22) has at most one positive root p_B^*. From this, (2.20) and (2.21), one obtains the wolutions w^* and r^* for each region; from (2.3) and (2.4) L^* and K^* for each region; from (2.18) and (2.19), $(B^S)^*$ and $(I^S)^*$ for each region. From (2.9) we then obtain $(B^D)^*$ for each region, and $(I^D(N))^*$ is computed from (2.11). All endogenous variables have been computed, and the solution is complete. ∎

The next theorem looks at the qualitative behavior of the solutions as an exogenous parameter changes. We study changes in the levels of the exogenous parameter $I^D(N)$. Since for each $I^D(N)$ there is a unique equilibrium, we are looking at a one-dimensional family of equilibria. Along this family of equilibria the level of exports of the South changes. We are interested in the relationship across equilibria between the level of exports $x_B^S(S)$ and the terms of trade, real wages and consumption of the South.

<u>Theorem 2.</u> <u>Consider a North-South economy as above where labor is very abundant in the South</u> ($\alpha(S)$ <u>large</u>). <u>An increase in the volume of exports</u> $x_B^S(S)$ <u>can lead to two different outcomes</u>:

(1) If at the initial equilibrium $c_2/D < 2w/p_B$ (e.g., <u>technologies are dual in the South</u>, i.e., D <u>is large</u>), <u>then an increase in exports leads to lower terms of trade for the South</u> (p_B), <u>lower export revenues</u> ($p_B x_B^S$), <u>lower real wages</u> (w/p_B), <u>and lower domestic consumption</u> (B^D) <u>in the South</u>.

(2) <u>When</u> $c_2/D > 2w/p_B$ (e.g., <u>technologies are more homogeneous, or real wages smaller</u>) <u>an increase in exports leads instead to higher export revenues</u> ($p_B x_B^S$), <u>higher terms of trade</u> (p_B), <u>higher real wages</u> (w/p_B) <u>and higher consumption in the South</u>.

Proof: Across equilibria, the following equation is satisfied:

$$X_B^S(S) = B^S(S) - B^D(S) = \frac{(c_2 L - a_2 K)}{D} - \left(\frac{wL + rK - \bar{I}^D(S)}{p_B}\right), \quad (2.23)$$

where the second equality comes from (2.18) and the Walras Law of footnote 2. Substituting L and K as functions of w and r, and w and r as functions of p_B (from (2.3), (2.4), (2.20) and (2.21)) we obtain

$$X_B^S(S) = \frac{\alpha c_1}{D^2 p_B} \left(c_2 - \frac{c_1}{p_B}\right) + \frac{\beta a_1}{D^2}\left(a_2 - \frac{a_1}{p_B}\right) + \frac{c_1 \bar{L} - a_1 \bar{K}}{D p_B} + \frac{\bar{I}^D(S)}{p_B} \quad (2.24)$$

We have therefore expressed the level of exports X_B^S as a function of the terms of trade p_B only. We can now consider the derivative across the manifold of equilibria of X_B^S with respect to p_B:

$$\frac{\partial X_B^S}{\partial p_B} = \frac{\alpha c_1}{D^2 p_B^2}\left(\frac{\alpha c_1}{p_B} - c_2\right) + \frac{\beta a_1^2}{D^2 p_B^2} + \frac{a_1 \bar{K} - c_1 \bar{L}}{pB^2} - \frac{\bar{I}^D(S)}{p_B}.$$

When α is large (and β small) $\partial X_B^S / \partial p_B$ has the sign of

$$\frac{1}{D}\left(\frac{2c_1}{p_B} - c_2\right) = \frac{c_2}{D} - \frac{2w}{p_B}$$

by (2.20). Therefore, when $c_2/D < 2w/p_B$ (case (1)), an increase in the volume of exports leads to a lower solution value p_B^*; when $c_2/D > 2w/p_B$ the opposite is true: as X_B^S rises, so does p_B^*. Our last task is to examine the impact of X_B^S on real wages, consumption and export revenues of the South.

It is easy to check that in case (1) ($c_2/D < 2w/p_B$) export revenues $p_B X_B^S$ <u>fall</u> as exports increase. This is because a higher level of X_B^S in this case leads to a lower p_B^* and from (2.20) to a lower $(w/p_B)^*$ and a higher r^*. By (2.3), (2.4) and (2.19), this means that $(I^S)^*$ increases. Since $\bar{I}^D(S)$ is a constant, X_I^D has decreased. Therefore by (2.11), $(p_B X_B^S)^*$ drops. The opposite effects happen in case (2): p_B increases, and so do $p_B X_B^S$ and w/p_B.

The effect of exports on domestic consumption is also easy to check. In case (1) ($c_2/D < 2w/p_B$), p_B^* drops, $(w/p_B)^*$ drops and r^* increases. From (2.3) and (2.4) L^* drops and K^* increases. This means that $(B^S)^*$ drops, from (2.18). Since

$$B^D = B^S - X_B^S ,$$

$(B^S)^*$ dropped and $(X_B^S)^*$ increases, $(B^D)^*$ must now be lower. The opposite happens when $c_2/D > 2w/p_B$. This completes the proof. ∎

It is of interest to examine what drives this result. For exports to increase, the difference between domestic supply and demand for basics must be wider. If an increases in prices p_B leads to proportionately higher increase in demand than in supply, then exports can only rise when prices drop. This is case (1): the expression $c_2/D - 2w/p_B$ measures the relative strength of the supply and demand responses as prices p_B change, c_2/D coming from the supply and $2w/p_B$ from demand. Notice that we do not refer here to standard supply and demand responses to prices when all other variables are kept constant. We refer instead to <u>general equilibrium</u> effects: as prices change everything else does too: employment, real wages and thus also income. Wage income wL increases with higher prices p_B, so that the general equilibrium effect of higher prices on demand is positive. However, the standard partial equilibrium effect of prices on demand, when all other variables are constant, is in general negative. From the proof of Theorem 2 we obtain:

<u>Corollary 3</u>. <u>The terms of trade of the South</u>, p_B, <u>are always positively associated with wage income</u> wL <u>and negatively associated with capital income</u> rK <u>in both regions</u>.

The above results show how a system of 26 equations in 26 variables can be solved analytically to yield qualitative results. In particular we showed the connection across equilibria between the volume of exports of the South X_B^S and main aggregate variables of the South: real wages, consumption levels, export revenues.

Theorem 2 shows how to answer questions (1) and (2) in the Introduction. The connection between exports x_B^S and their price is given by equation (2.24) in Theorem 2. The distribution of income, measured here as the relative level of wages w (or real wages w/p_B) and the return on capital r, is in turn changing with exports through the relationships (2.20) and (2.21), as pointed out in Corollary 3. Finally, Theorem 2 shows when a country should export more (α large, $c_2/D > 2w/p_B$) or less (α large, $c_2/D < 2w/p_B$).

The next section will focus on the responses of the importing economy to exogenous changes in resource prices.

3. A Monopolistic Resource Exporter

A. Specification of the Model

The previous model dealt with trade in which prices adjust so as to clear the markets. No single agent influences prices; the economic agents are, therefore, "competitive." This section will deal instead with a case where one region fixes the price of the resource it exports. This agent is therefore called a monopolist.

The model of the last section will now be modified to include one more input of production, oil. The price of oil is now an exogenously set parameter. We shall focus on the effects of changes in oil prices on the economy of the importer. For this purpose we concentrate on modelling the economy of the North. This model was introduced in Chichilnisky (1981b).

The North produces basics (B) and industrial goods (I) using capital (K), labor (L) and oil (θ):

$$B^S = \min(L^B/a_1,\ \theta^B/b_1,\ K^B/c_1)$$

$$I^S = \min(L^I/a_2,\ \theta^I/b_2,\ K^I/c_2)\ .$$

The "zero profit" price equations, as explained in the previous sections, are

$$p_B = a_1 w + b_1 p_o + c_1 r p_I \tag{3.1}$$

$$p_I = a_2 w + b_2 p_o + c_2 r p_I \tag{3.2}$$

where $p_I r$ denotes here the "user's cost" of capital. Supplies of labor and capital depend on their rewards:

$$L^S = \alpha w/p_B,\qquad \alpha > 0 \tag{3.3}$$

$$K^S = \beta r,\qquad \beta > 0\ . \tag{3.4}$$

Next we formulate a demand equation, which substitutes for the exogenously set industrial demand I^D of the previous section:

$$p_B B^D = wL , \qquad (3.5)$$

i.e., wage income is spent on basics. This relation is not necessary to prove the results, but it simplifies considerably the computations.

The market clearing conditions are

$$L^D = L^B + L^I = a_1 B^S + a_2 I^S \qquad (3.6)$$

$$K^D = K^B + K^I = c_1 B^S + c_2 I^S \qquad (3.7)$$

$$\theta^D = \theta^B + \theta^I = b_1 B^S + b_2 I^S \qquad (3.8)$$

$$K^S = K^D \qquad (3.9)$$

$$L^D = L^S \qquad (3.10)$$

$$\theta^D = \theta^S \qquad (3.11)$$

$$B^D = B^S \quad \text{(B is not traded internationally)} \qquad (3.12)$$

$$I^D + X_I^S = I^S \quad (X_I^S \text{ are exports of I}) \qquad (3.13)$$

$$p_I X_I^S = p_o X_\theta^D \quad \begin{array}{l}\text{(value of exports of I equals value of imports} \\ \text{of oil)}\end{array} \qquad (3.14)$$

$$\theta^D = X_\theta^D \quad \text{(the North imports the oil it consumes)} \qquad (3.15)$$

and basics are the numeraire

$$p_B = 1 . \qquad (3.16)$$

As before, the Walras Law or national income identity is identically satisfied when (3.1) through (3.15) hold: the value of domestic demand equals the value of domestic income

$$p_B B^D + p_I I^D = wL + rK. \tag{3.17}$$

The <u>monopolistic oil-exporter model</u> is specified as follows. Its exogenous parameters are the technical coefficients a_1, a_2, b_1, b_2, c_1, c_2; the supply parameters α and β; and the price of oil, p_o.

There are sixteen endogenous variables: p_B, p_I, r, w, L^D, L^S, K^D, K^S, B^D, B^S, I^D, I^S, X_I^S, θ^D, X_θ^D, θ^S; and sixteen independent equations (3.1) through (3.16). Therefore there are, as before, as many variables as independent equations, and the solutions will generally be (locally) unique. The following results explore the effects of changes in the exogenous price of oil p_o on the fifteen endogenous variables of the North. To simplify computations we now assume the following stylized conditions:

I. $a_1 b_2 - a_2 b_1 = M > 0$, i.e., B is relatively more labor-intensive and I more oil-intensive;

II. $c_1 = 0$, i.e., B requires no capital income (this condition is not necessary, but simplifies greatly the computations);

III. b_1 is small, i.e., B requires relatively few oil inputs.

<u>Proposition 4.</u> <u>The monopolistic oil-exporter model has at most one solution for each price of oil p_o. This solution can be explicitly computed by solving one equation.</u>

<u>Proof.</u> Equations (3.6) and (3.7) imply

$$B^S = (c_2 L - a_2 K)/D \tag{3.18}$$

and

$$I^S = (a_1 K - c_1 L)/D \tag{3.19}$$

where $D = a_1 c_2 - a_2 c_1$.

The price equations (3.1) and (3.2) yield

$$w = \frac{(p_B - b_1 p_0)c_2 - (p_I - b_2 p_0)c_1}{D} \quad (3.20)$$

and

$$r = \frac{a_1(p_I - b_2 p_0) - a_2(p_B - b_1 p_0)}{D} \quad (3.21)$$

Substituting in (3.18) K and L from (3.3) and (3.4) and then w and r from (3.20) and (3.21) one obtains

$$B^S = \frac{(c_2 \alpha w - a_2 \beta r)}{D} = \frac{\alpha c_2}{D^2}(c_2 + p_0 N - c_1 p_I) + \frac{\beta a_2}{D^2}\left(\frac{p_0}{p_I} M + \frac{a_2}{p_I} - a_1\right) \quad (3.22)$$

where $M = a_1 b_2 - a_2 b_1$ and $N = c_1 b_2 - b_1 c_2$.

From $B^D = wL$ and $B^S = B^D$, one obtains, when $p_B = 1$,

$$\alpha c_2(c_2 + p_0 N - c_1 p_I) + \beta a_2\left(\frac{p_0}{p_I} M + \frac{a_2}{p_I} - a_1\right)$$

$$= \alpha\left[(1 - b_1 p_0)c_2 - (p_I - b_2 p_0)c_1\right]^2 . \quad (3.23)$$

When $c_1 = 0$, this yields an explicit relation between the price of oil and the price of industrial goods

$$p_I = \frac{a_2 + p_0 M}{\gamma b_1 p_0 (b_1 p_0 - 1) + a_1} \quad (3.24)$$

where

$$\gamma = \frac{\alpha}{\beta} \frac{c_2^2}{a_2} .$$

From equation (3.24), there exists at most one solution p_I^* for each p_0. Given p_I^*, we obtain all other endogenous variables as follows: w^* obtains from (3.20), r^* from (3.21), L^* and K^* from (3.3) and (3.4), $(B^S)^*$ and $(I^S)^*$ from (3.18) and (3.19) and $(B^D)^*$ from (3.5). $(\theta^D)^* = (X_\theta^D)^*$ obtains from (3.8), and therefore from (3.14) one obtains $(X_I^S)^*$. All endogenous variables are uniquely determined. This completes the proof. ∎

The next results explore the comparative statics properties of changes in the price of oil, p_o.

B. Comparative Statics Results

Theorem 5. Consider a monopolistic exporter model as in Proposition 4. An increase in the (monopolistically set) price of oil has the following effects: If the initial price of oil is low, increases in this price benefit the exporter; the volume of industrial goods traded in exchange for oil X_I^S increases. However, after a price $\bar{p}_o$ has been reached, the volume of industrial exports X_I^S decreases with further increases in the price of oil; the monopolist loses from increasing the price p_o. There are two cases:

(A) $\alpha c_2^2 \leq 2\beta a_1 a_2$. In this case $\bar{p}_o = 1/2b_1$, and the rate of profit r rises and falls with the level of exports X_I^S, see Figure 1.

(B) $\beta c_2^2 > 2\beta a_1 a_2$. In this case

$$\bar{p}_o = \frac{1}{2b_1}\left(1 - \frac{(\gamma - 2a_1)^{1/2}}{\gamma}\right).$$

Between $\bar{p}_o$ and $1/2b_1$, r increases and X_i^S decreases. Between $1/2b_1$ and $(1/b_1) - \bar{p}_o$ the opposite happens: X_I^S increases and r drops, see Figure 2.

Therefore in case (A) the oil exporter and the oil importer can both benefit from an increase in p_o when $p_o < 1/2b_1$ and from a decrease in p_o when $p_o > 1/2b_1$. This is provided that their aim is to maximize returns on capital (r) and the volume of industrial exports traded for oil (X_I^S) respectively.

In case (B), there are areas where both traders benefit from a price increase $(p_o < \bar{p}_o)$ or from a price drop $(p_o > (1/b_1) - \bar{p}_o)$. Outside those areas, the traders have conflicting interests.

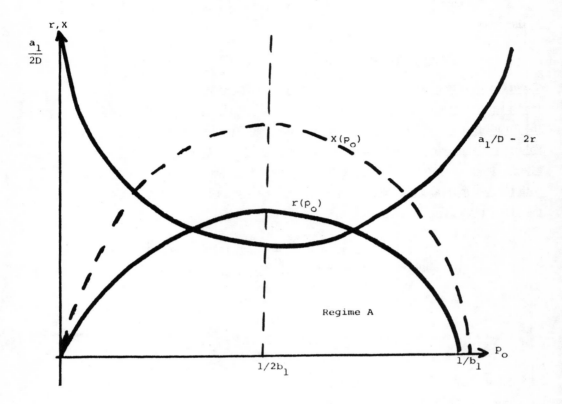

Figure 1. Case 1: $\alpha c_2^2 > 2\beta a_1 a_2$, i.e., r is always bounded below $a_1/2D$. In this case, as the price of oil p_o increases, the volume of industrial exports $X(p_o)$, and the rate of return on capital $r(p_o)$ increase and decrease together.

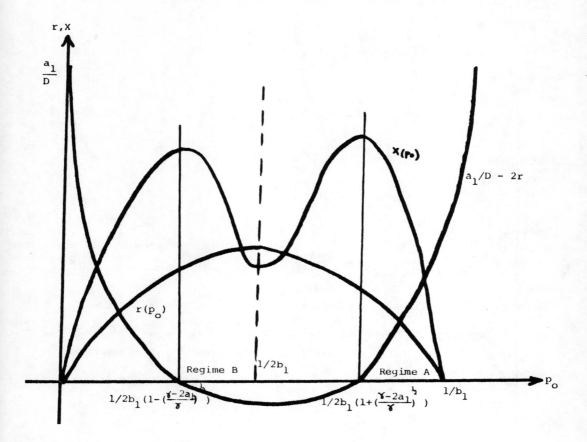

Figure 2. Case 2: In this case, the rate of return on capital r and the level of industrial exports traded for oil X, do not move in the same direction in regime B. There are therefore regions with community of interests for the traders, and others where there are conflicting interests.

Proof. By (3.21) r is a function of p_I and p_o. By (3.24) $p_I = p_I(p_o)$. These two relations yield

$$r = \frac{\alpha c_2 b_1}{\beta a_1 a_2} (p_o - b_1 p_o^2) , \qquad (3.25)$$

so that

$$\frac{dr}{dp_o} = \frac{\alpha c_2 b_1}{\beta a_1 a_2} (1 - 2p_o b_1) .$$

Also, since $X_I^S = I^S - I^D$, from (3.19), (3.5) and (3.17)

$$X_I^S = \beta \left(\frac{a_1 r}{D} - r^2 \right) , \qquad (3.26)$$

so that

$$\frac{dX_I^S}{dp_o} = \beta \left(\frac{a_1}{D} - 2r \right) \frac{dr}{dp_o} . \qquad (3.27)$$

An examination of equations (3.25) through (3.27) suffices to prove the results.

Equation (3.25) means that r is a quadratic function of p_o, $r_0 = 0$ when $p_o = 0$, $r_0 = 0$ when p_o attains its maximum feasible value $1/b_1$. The maximum value of r,

$$\frac{1}{4} \frac{\alpha c_2^2}{\beta a_1 a_2} ,$$

is attained when $p_o = 1/2b_1$. From equation (3.27) when $a_1/D > 2 \max r$, then dX_I^S/dp_o has the same sign as dr/dp_o. Now, $a_1/D > 2 \max r$ if and only if

$$\frac{a_1}{D} > \frac{1}{2} \frac{\alpha c_2}{\beta a_1 a_2} ,$$

i.e., $\alpha c_2^2 < 2\beta a_1 a_2$. This is case (A), see Figure 1. Case (B) is when $\alpha c_2^2 > 2\beta a_1 a_2$. In this case dX_I^S/dp_o has the opposite sign to dr/dp_o when $p_o^1 \leqslant p_o \leqslant p_o^2$, for

$$p_o^1 = \frac{1}{2b_1} \left(1 - \left(\frac{\gamma - 2a}{\gamma} \right)^{1/2} \right) , \quad p_o^2 = \frac{1}{2b_1} \left(1 + \left(\frac{\gamma - 2a}{\gamma} \right)^{1/2} \right)$$

as in Figure 2. This completes the proof. ∎

The previous theorem does not explain the impact of oil prices on the GDP of the importing region. This is measured as the total value of domestic output minus the value of imported inputs:

$$GDP = Y = p_B B + p_I I - p_o \theta .$$

Theorem 6. <u>For very low prices of oil, an increase in p_o lowers the GDP of the importer. However, after a price p_o has been reached, further increases increase the GDP, up to p_o. From this latter point on, the effect of oil prices on GDP is again negative.</u>

Proof: From (3.3), (3.4), (3.5) and (3.17),

$$Y = \alpha w^2 + p_I \beta r^2 , \qquad (3.28)$$

so that

$$\frac{dY}{dp_o} = \alpha 2w \frac{dw}{dp_o} + \beta \left(r^2 + \frac{dp_I}{dp_o} + p_I 2r \frac{dr}{dp_o} \right) . \qquad (3.29)$$

Note that from (3.20)

$$\frac{dw}{dp_o} = -b_1 c_2 / D < 0 \qquad (3.30)$$

Also, since $p_B B^D = wL = \alpha w^2$,

$$\frac{dB^D}{dp_o} = \alpha 2w \frac{dw}{dp_o} < 0 .$$

Furthermore, $r = 0$ when $p_o = 0$. Therefore (3.29) implies that for small values of p_o, Y is a decreasing function of p_o, $dY/dp_o < 0$. However, since $dr/dp_o > 0$ for $p_o \leq 1/2b_1$ and $b_1 \sim 0$, when p_o exceeds a small value (denoted p_o^3), Y is an increasing function of the price of oil. This is due to the fact that as p_o increases, the value of demand for I, p_I^D increases, since dr/dp_o and $dp_I/dp_o \geq 0$ for

$$p_o \leq \frac{1}{2b_1} \left(1 + \left(\frac{\gamma - 2}{\gamma} \right) \right)^{1/2} .$$

Since b_1 is rather small, the increase in p_I^D within Y exceeds the decrease in p_B^D, so that Y increases with the price of oil. Figure 3 illustrates this result from a computer simulation of the model.

Finally, when p_o exceeds a value p_o^4, dY/dp_o turns negative again, because $r = 0$ when p_o assumes its maximum value $1/b_1$, $dr/dp_o < 0$ for $p_o > 1/2b_1$, $dB^D/dp_o < 0$ and rdp_I/dp_o is bounded above by

$$-\frac{M\alpha c_2 b_1}{\beta a_1 a_2} p_o$$

since

$$\frac{dp_I}{dp_o} = \frac{M(a_1 + b_1 \gamma p_o (b_1 p_o - 1)) - (a_2 + p_o M)\gamma b_1 (2 p_o b_1 - 1)}{(a_1 + \gamma b_1 p_o (p_o b_1 - 1))^2},$$

for $\gamma = \frac{\alpha}{\beta} c_2^2 / a_2$. This completes the proof. ∎

The last two theorems answer questions 3, 4 and 5 of the Introduction. They show how the price of an imported resource affects prices, output and exports of the importing region. They show how a monopolistic resource exporter should take into account the effect of its prices on the importing economy: when to raise and when to decrease resource prices. Finally, they demonstrate cases where the importer and the exporter have a conflict of interest, and others where both could benefit from a rise, or a fall, in resource prices.

C. <u>Computer Simulation</u>

The following is a computer simulation of this model. The following values of the exogenous parameters are given:

Exogenous Data

$\alpha = 1.00$	$b_1 = 0.10$
$\beta = 2.00$	$b_2 = 0.20$
$a_1 = 0.30$	$c_1 = 0.001$
$a_2 = 0.20$	$c_2 = 0.60$

As explained in Proposition 4, a solution of the model is computed by resolving one equation in p_o, equation (3.23), and computing all other endogenous variables from p_o. The following table reproduces the numerical results for different values of the price of oil, p_o. Figures 3 and 4 reproduce graphically these results. Figure 3 illustrates the response of GDP = Y, the rate of return on capital r, the volume of industrial exports X, wages w and the price of industrial exports, as the price of oil varies. Figure 4 illustrates the connection between oil prices p_o and oil exports θ^S, across equilibria of the North-South economy.

TABLE 1

Exogenous p_o	p_I	r	w	Endogenous $p_I^* X$	Y	θ^S
0.0	0.668901	0.5574d-02	3.3332	0.337150d-16	11.1111	1.01
0.5	0.858613	0.243452	3.16597	0.580648	10.1251	1.16
1.0	1.10132	0.456505	2.99832	1.20033	9.44897	1.20
1.5	1.41239	0.644688	2.83030	1.84227	9.18463	1.228
2.0	1.80953	1.807918	2.66179	2.48953	9.44742	1.244
2.5	2.30945	0.946045	2.49272	3.12497	10.3476	1.249
3.0	2.91912	1.05883	2.32303	3.73121	11.9419	1.243
3.5	3.61864	1.14595	2.15284	4.29064	14.1388	1.225
4.0	4.33894	1.20707	1.98254	4.78579	16.5742	1.196
4.5	4.95548	1.24203	1.81282	5.20000	18.5754	1.155
5.0	5.33000	1.25113	1.64444	5.51801	19.3906	1.103
5.5	5.38881	1.23504	1.47782	5.72565	18.6234	1.041
6.0	5.16262	1.19450	1.31278	5.80860	16.4559	0.968
6.5	4.75047	1.12996	1.14887	5.75144	13.4507	0.884
7.0	4.25777	1.04152	0.985218	5.53761	10.2080	0.791
7.5	3.76058	0.929042	0.821688	5.14990	7.16683	0.686
8.0	3.30079	0.792336	0.657949	4.57096	4.57735	0.571
8.5	2.89531	0.631209	0.493908	3.78351	2.55107	0.445
9.0	2.54664	0.445510	0.329551	2.77048	1.11951	0.307
9.5	2.25047	0.235131	0.164903	1.51491	0.276035	0.159
10.0	2.00000	0.385923d-01	0.	0.257568d-15	0.595746	0.000017

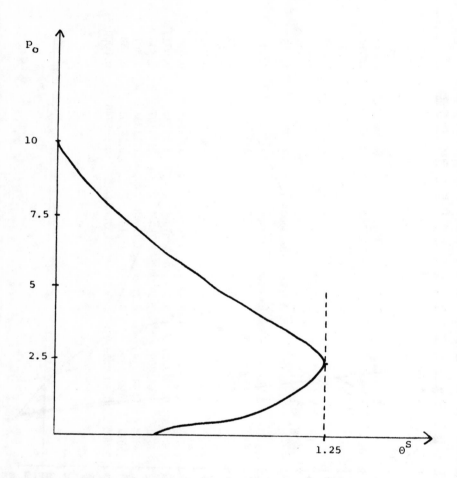

Figure 3. The volume of oil exports and the price of oil across equilibria.

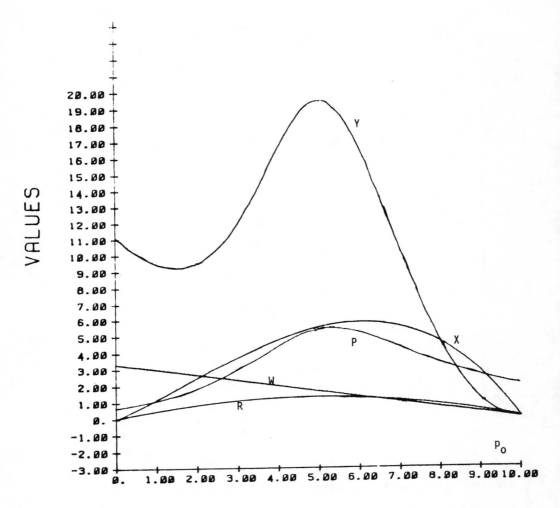

Figure 4. Results from a simulation: note that the GDP, Y, of the oil importer rises and then falls with increases in the price of oil, p_o.

4. Debt and Resource Exports

In this section we extend the North-South model to study the impact of debt on the resource-importing and resource exporting economics. We allow here for an imbalance in the trade account, which is matched by an inflow of overseas investment or a <u>financial transfer</u>. This imbalance represents the debt owed to foreigners, and is directed towards the expansion of oil supplied. Except for the wedge between export revenues and import costs, which represents the debt, the model is consistent with a standard competitive general equilibrium specification. The model was introduced in Chichilnisky, Heal and McLeod (1984).

The introduction of the debt wedge changes the main relations in the model: the operation of Walras' Law or the national income identity in both countries is altered. Overseas transfers lead to changes in oil supplies and consequently most variables adjust. As the debt increases, a new equilibrium emerges with different prices and levels of imports and exports. There are also changes in all domestic variables in both South and North: real wages, profits, domestic use of industrial and consumption goods, and employment of the factors labor, capital and oil. This allows us to trace the impact of the debt on the major macro variables of the two countries. The model could also be used to examine the impact of rescheduling, i.e., repaying the debt over a different time period, or of repaying it at a different rate of interest.

Following the macroeconomic impact analysis, two main questions emerge. The first is, who benefits and who loses from the accumulation of debt; and the second is, whether there exist debt-management policies that could make both countries better off, after taking fully into account the recycling effect of borrowing funds on imports from the lender.

The interest of the results lies in part in their simplicity and in part in the fact that they account for the impact of the debt on all markets simultaneously. Fairly simple analytical solutions are obtained to the rather complex questions posed. These are obtained, of course, at the cost of somewhat stylized assumptions.

We describe conditions under which increasing the debt leads the country to export more oil. In certain cases, this leads to lower prices of oil, lower volumes of industrial imports, lower real wages, and higher profits in the oil-exporting country. In other cases, the results are reversed, and real wages, consumption, and terms of trade all improve in the exporting country. The outcome depends on the technologies of the South an on the initial prices.

We also examine conditions under which the economy of the North actually benefits in macroeconomic terms from its loan to the South: because of lower oil prices, the consumption of both goods increases in the North when the transfer or loan increases. This occurs because the transfer leads to better terms of trade in the North, and because its production system is relatively homogeneous. This result is reminiscent of the argument that British investment overseas in the nineteenth century benefited the country by developing overseas supplies of food and raw material, thus making these supplies more elastic, keeping down prices, and improving the UK's terms of trade. Essentially we are specifying here conditions for overseas investment in material supplies to benefit the investing country even before any financial returns are paid, or in the case of a loan, before the loan is repaid.

A. Specification of the Model

There are two regions, the North and the South. Each produces two goods, denoted B and I, with three factors of production, capital, K; labor, L; and oil, θ. The South exports an input, oil, in exchange for a good, the "industrial" good I. The "basic" good B is produced domestically and not traded internationally.

We first specify the model for one region, namely the South. In what follows, the subscripts S and D will be used to denote supply and demand, and the superscripts N and S to denote variables or parameters referring to the North and South, respectively. All variables or parameters without a superscript refer to the South. The superscripts B and I after a factor (e.g., L^B, K^I) denote the amount of that factor used in sector B or I, respectively.

The basic good is produced according to the relation

$$B_S = \min\left[L^B/a_1, \theta^B/b_1, K^B/c_1\right] \tag{4.1}$$

and the industrial good according to

$$I_S = \min\left[L^I/a_2, \theta^I/b_2, K^I/c_2\right] \tag{4.2}$$

Labor and capital supplies are responsive to their rewards:

$$L_S = \alpha w/p_B, \quad \alpha > 0 \tag{4.3}$$

where w is the wage and p_B the price of B, and

$$K_S = \beta r, \quad \beta > 0 \tag{4.4}$$

where r is the rate of profit. p_I and p_θ will stand for the prices of industrial goods and of oil, respectively. The demand for B derives from wage income

$$p_B B_D = wL \tag{4.5}$$

The South produces oil (within given bounds) without using either domestic capital or labor. We shall assume that it uses the overseas borrowing or financial transfer FT to increase its oil supplies

$$\theta_S = \theta_S(FT), \quad \partial\theta_S/\partial FT > 0 \tag{4.6}$$

This completes the behavioral specification for the South.

The equilibrium conditions for the South are:

$$B_S = B_D \tag{4.7}$$

where B is not traded internationally.

$$I_D = I_S + M_I^S \tag{4.8}$$

where M_I^S denotes the South's imports of I.

$$\theta_S^S = \theta_D + X_\theta^S \qquad (4.9)$$

where X_θ^S denotes oil exports by the South,

$$K_S = K_D \qquad (4.10)$$

$$L_S = L_D \qquad (4.11)$$

$$L_D = B_S a_1 + I_S a_2 \qquad (4.12)$$

$$K_D = B_S c_1 + I_S c_2 \qquad (4.13)$$

$$\theta_D = B_S b_1 + I_S b_2 \qquad (4.14)$$

and the payments condition

$$p_\theta X_\theta^S = p_I M_I^S - FT, \qquad (4.15)$$

where FT denotes financial transfers.

Note that FT could be either positive or negative, depending on the relative magnitudes of the debt service and the financial credit. However, the effect of a transfer (FT positive) may not be symmetric with that of a repayment (FT negative), because of the irreversibility of the investment in the oil sector: we assume that the financial transfer FT is used to purchase industrial goods to augment the supply of oil. This means that the new industrial investment in the oil sector is paid for by foreign loans. Hence, oil supplies θ_S change as the debt level changes; the debt is assumed to increase with increases in the level of the transfer (FT positive). The balance-of-payments condition (4.15) is that imports of industrial goods exceed export revenues by FT. As the demand for the basic good B comes entirely from wage income (4.5), the national income identity ((4.16) below) implies that the demand for industrial goods comes from the profit income rK, oil revenues $p_\theta X_\theta^S$, and the borrowing FT, with the last of these going to the oil sector. In the North we make a corresponding assumption, namely that the financial transfer to the South is taken from income that would otherwise have purchased industrial goods, so that the North's demand for industrial goods is rK - FT.

In an equilibrium situation, Walras' Law or the national income identity of the South is always satisfied (see Chichilnisky, 1981a), i.e.,

$$p_B B_D - p_I I_D = wL + rK + p_\theta \theta + FT \qquad (4.16)$$

where $\theta = \theta_S$ is, as in (4.6), a function of FT. Equation (4.16) can also be rewritten as

$$p_B B_S + p_I(I_S + M_I^S) = wL + rK + p_\theta(\theta_D + X_\theta^S) + NF . \qquad (4.16)'$$

This completes the specification of the South.

The model of the North consists of the same 15 equations, but with possibly different parameters α, β, a_1, a_2, b_1, b_2, c_1, c_2. The following equation now substitutes for equation (4.6) in the South:

$$\theta_S = 0 , \qquad (4.6)'$$

Of course, the equations corresponding to (4.8) and (4.9) reflect the fact that the North imports oil and exports industrial goods. In a world trade equilibrium the prices of the traded goods must be equal:

$$p_\theta^S = p_\theta^N \qquad (4.17)$$

$$p_I^S = p_I^N ; \qquad (4.18)$$

and traded quantities must also match:

$$X_\theta^S = M_\theta^N \qquad (4.19)$$

$$X_I^N = M_I^S \qquad (4.20)$$

where X_I^N and M_θ^N represent, respectively, the North's exports of I and imports of oil. There are therefore two sets of eight exogenous parameters each, one set for the North and the other for the South. Each set contains α, β, a_1, a_2, b_1, b_2, c_1 and c_2. These parameters are generally different in the two regions.

We make certain stylized assumptions to simplify computations: α is large in the South and relatively smaller in the North, indicating that labor is more "abundant" in the South. The corresponding parameters for capital exhibits the opposite behavior: β is large in the North than in the South. We shall also assume that c_1 is small in the South, i.e., the production of basic goods uses little capital, and a_2 is small in the North, i.e., Northern industry uses little labor.

There are a <u>total of</u> 33 <u>independent equations for the complete North-South system</u>: thirty correspond to two sets of (4.1) through (4.15), one set for each region, and three equations arise from the international trade conditions (4.17) through (4.20), since of these four, as usual, only three are linearly independent. There are 17 endogenously-determined variables each in the North and in the South: p_I, p_θ, p_B, w, r, L_S, L_D, K_S, K_D, B_S, B_D, I_S, I_D, M_I^S, θ_S, θ_D and X_θ^S. Finally, we have the transfer FT, making a <u>total of</u> 35 <u>endogenous variables for the complete North-South system</u>. We therefore have 33 equations in 35 unknowns. When we choose the **numeraire** ($p_\theta = 1$) an equilibrium is determined up to one variable. If we fix exogenously one variable, the equilibrium is (locally) unique. We choose this variable to be the value of the transfer FT. The transfer or loan thus becomes a policy variable. In the following we show how to compute explicitly a solution to the model, i.e., a value for each of the endogenous variables for each value of FT. In particular, we show that by successive substitutions the more important properties of the model can be obtained from the study of a <u>single equation</u>, giving an implicit relationship between the financial transfer FT and the relative prices of industrial goods and oil.

There are a number of determinants whose signs are important in the following, and which determine factor intensities in the different sectors. We have the following technical input-output coefficients:

$$\begin{bmatrix} a_1 & b_1 & c_1 \\ a_2 & b_2 & c_2 \end{bmatrix}$$

in each region. The determinants to be used are:

$$D = a_1 c_2 - a_2 c_1, \quad M = c_1 b_2 - b_1 c_2, \quad Q = a_2 b_1 - a_1 b_2$$

The assumptions are:

$$D^N > 0, \quad D^S > 0, \quad M^S < 0, \quad Q^N < 0.$$

The positivity of the determinant D implies that the basic goods sector is relatively more labor intensive and the industrial goods sector relatively more capital intensive. The assumption (made above) that the basic goods sector uses very little capital in the South implies that c_1^S is small and therefore that $M^S < 0$. The industrial goods sector in the North was assumed to use little labor: hence a_2^N is small and $Q^N < 0$.

In order to solve the model we consider first the equation equating oil exported with oil imported:

$$X_\theta^S = M_\theta^N . \tag{4.21}$$

In view of (4.6), (4.9), and (4.6)', this equals

$$\theta_S(FT) - \theta_D = \theta_D^N , \tag{4.22}$$

where the left-hand-side variables are from the South. From (4.14), (4.12) and (4.13)

$$\theta_D = \frac{b_1}{D}(c_2 L - a_2 K) + \frac{b_2}{D}(a_1 K - c_1 L) \tag{4.23}$$

$$= -\frac{\alpha}{D}\frac{w}{p_B} M - \frac{\beta r}{D} Q ,$$

where

$$M = c_1 b_2 - b_1 c_2 , \qquad Q = a_2 b_1 - a_1 b_2 .$$

Therefore, we may rewrite (4.22) as

$$\theta_S(FT) - \frac{\alpha}{D}\frac{w}{p_B} M - \frac{\beta r}{D} Q = \frac{a^N}{D^N}(w/p_B)^N M^N - \frac{\beta^N r^N Q^N}{D^N} . \tag{4.24}$$

Equation (4.24) is therefore an implicit equation in five variables, which we denote

$$\phi\left(FT,\ r,\ w/p_B,\ r^N,\ (w/p_B)^N\right) = 0 \quad . \tag{4.25}$$

Our next step is to write the rate of profit r and the wage w/p_B in the two regions as functions of the prices of basic and industrial goods, p_B and p_I. Recall that oil is the numeraire ($p_\theta = 1$). From the production functions (4.1) and (4.2) we obtain the associated competitive price equations

$$p_B = a_1 w + b_1 p_\theta + c_1 r \ , \qquad p_I = a_2 w + b_2 p_\theta + c_2 r \ , \tag{4.26}$$

i.e.,

$$\begin{bmatrix} p_B - b_1 \\ p_I - b_2 \end{bmatrix} = \begin{bmatrix} a_1 & c_1 \\ a_2 & c_2 \end{bmatrix} \begin{bmatrix} w \\ r \end{bmatrix}$$

since $p_\theta = 1$. We therefore obtain the factor-commodity price relations

$$w = \frac{c_2 p_B - c_1 p_I + M}{D}$$

$$w/p_B = \frac{(p_B - b_1) c_2 + (b_2 - p_I) c_1}{D p_B} \tag{4.27}$$

$$r = \frac{(b_1 - p_B) a_2 + a_1 (p_I - b_2)}{D} = \frac{a_1 p_I - a_2 p_B + Q}{D} \quad .$$

Substituting w/p_B and r from (4.27) into (4.25), we obtain a new implicit function, in four, rather than five, variables:

$$\psi(FT,\ p_I,\ p_B^N,\ p_B^S) = 0 \quad . \tag{4.28}$$

Recall that p_B^N may be different from p_B^S because B is not traded internationally. The last step is to substitute p_B^N and p_B^S as functions of p_I into (4.28). This will lead to an implicit function in two variables

$$\chi(FT,\ p_I) = 0 \quad . \tag{4.29}$$

Since FT is an exogenously given parameter, (4.29) is an analytic solution to the model: from (4.29) we may compute the equilibrium level of industrial prices $p_I^*(FT)$. It is easy to check that once p_I^* is known, we may solve for the equilibrium values of all other endogenous variables. This will be explained below.

Now, in order to obtain $p_B = p_B(p_I)$, we use another market-clearing condition, this time in the B-market:

$$B_S = B_D . \qquad (4.30)$$

From (4.12) and (4.13) this can be written as

$$\frac{c_2 L - a_2 K}{D} = \frac{w}{p_B} L \qquad (4.31)$$

or

$$a \frac{c_2}{D} \frac{w}{p_B} - \frac{\beta a_2 r}{D} - a(w/p_B)^2 = 0$$

from which we obtain

$$\frac{w}{p_B} = \frac{c_2}{D} \pm \left(\frac{c_2^2}{4D^2} - \frac{\beta a_2 r}{Da}\right)^{1/2} , \qquad (4.32)$$

a two-branched function relating w/p_B and r. The different parameter values will determine which is the appropriate branch in (4.32).

Using again the factor-commodity price relations, (4.32) yields an implicit relation between p_B and p_I as desired:

$$\frac{c_2}{aDp_B} \pm \frac{1}{p_B} \left(\frac{c_2^2}{4D^2} - \frac{\beta a_2}{D^2 a}(Q - a_2 p_B + a_1 p_I)\right)^{1/2} - \frac{M}{Dp_B} \frac{c_2}{D} + \frac{c_1 p_I}{Dp_B} = 0 . \qquad (4.33)$$

Substituting (4.33) into (4.28), we obtain the desired relation (4.29) between FT and p_I,

$$\chi(FT, p_I) = 0 .$$

From (4.29) we may then compute $p_I^* = p_I^*(FT)$. From (4.33) we obtain $p_B^*(N)$ and $p_B^*(S)$, and from these three equilibrium prices we obtain the equilibrium rates of profit $r^*(N)$ and $r^*(S)$, and of real wages, $(w/p_B)^*(N)$ and $(w/p_B)^*(S)$. From these we obtain supply of labor and capital in the North and the South, and using the inversion of (4.32) and (4.33) we obtain the output of B and I in both regions. From the national income identity we may compute demand for I in the South, which determines imports from the North, and from exports of oil from the South. From (4.14) we obtain oil demanded in the South, thus completing the computation of the equilibrium.

B. <u>Main Results: Trade and Debt</u>

This section studies the impact of a change in the net transfer FT on the economies of the North and the South. Before going on to the algebra, it seems useful to explain the economics of this impact.

An increase in the transfer FT increases oil supplies θ_S, since the South invests borrowed funds in expanding the oil sector. At the new equilibrium, corresponding to higher FT, the total amount of oil utilized in the North and in the South therefore increases. This in turn alters the supplies of both goods in each region, possibly in different proportions. The composition of the product changes in both regions.

The changes in supplies lead to new equilibrium prices for the two goods. The prices of the factors labor and capital also change as relatively more or less labor and capital are employed. This implies that total income in the North and in the South are different at the new equilibrium. The results in this section give simple sufficient conditions for determining the signs of each of these effects.

The first theorem gives conditions under which an increase in oil supplies decreases the price of oil with respect to that of the industrial good. While it is intuitively plausible that the price of oil should drop as supplies increase, this is not always true. The second theorem gives conditions under which the relative price of oil <u>increases</u> as the transfer increases oil supplies. Whether one or the other result obtains depends on the relative strength of supply and demand effects, and the general equilibrium solutions trace this in detail. The results are obtained from various assumptions on technologies and initial prices.

The next step is to explore the general equilibrium impacts of an increase in the relative price of industrial goods. The rate of profit rises both in the North and in the South. In the North, the rate of profit and the real wage move together, because the North's economy is rather homogeneous. Therefore, both wage and profit income increase in the North, and we show that there is also an increase in the consumption of both goods, even allowing for the loss of national income due to the transfer. All this occurs because the transfer has improved significantly the North's terms of trade.

In the South, because of the rather different technologies in the two sectors, the real wage moves in the opposite direction to the rate of profit. The transfer increases oil supplies and oil exports, but oil revenues in terms of industrial goods imported are reduced. Wage income and domestic consumption of basics decrease as well. If one sought to improve wage income without negatively affecting industrial consumption in the South, the economy of the South would have to be made more homogeneous.

The second theorem explores a different set of assumptions, and arrives at rather different conclusions. Now the transfer increases oil supplies, but it also increases the relative price of oil with respect to industrial goods. As the terms of trade of the South improve, its macro variables react differently, and so do the variables in the North. The conditions under which one or the other result obtains are therefore quite relevant for policy, and should be determined empirically. The simulations for the case of Mexico in Chichilnisky, Heal and McLeod (1984) are a first step in this direction.

A factor that plays an important role in determining the results of an increase in the transfer FT is the sign of the expression

$$\Delta = \left[c_2/D - 2w/p_B \right]$$

where D is the determinant of the matrix

$$\begin{bmatrix} a_1 & c_1 \\ a_2 & c_2 \end{bmatrix}.$$

The role and interpretation of this term have been discussed elsewhere (Chichilnisky, 1981a,b). Basically, the sign of this expression determines whether income effects will dominate price effects, so that increases in supplies will be proportionately larger or smaller than increases in demand as prices change. We refer to an economy as <u>dual</u> if $c_2/D < 2w/p_B$, since a large D would have this interpretation. Conversely, the economy is <u>homogeneous</u> if $c_2/D > 2w/p_B$. It should be noted that this condition can be written so as to be independent of the particular units of measurement used.

Theorem 7. Consider a North-South economy as defined above. <u>Assume the economy of the North to be homogeneous</u> ($c_2/D > 2w/p_B$) <u>and that of the South to be dual</u> ($c_2/D < 2w/p_B$). <u>Suppose that at the initial equilibrium the price of industrial goods and the rate of profit are relatively high in the North</u> ($p_I < b_2$ <u>and</u> $2r > a_1/D$). <u>Labor is relatively abundant in the South</u> (α <u>large</u>) <u>and capital relatively abundant in the North</u> (β <u>large</u>). <u>In this case an increase in the transfer FT to the South has the following consequences</u>:

(i) <u>Oil supplies and oil exports increase in the South</u>.

(ii) <u>The North exports, and the South imports, fewer industrial goods. However, the terms of trade move in favor of the North</u> (p_I <u>increases</u>) <u>so much that its export revenues rise. There is a corresponding fall in oil export revenues of the South denominated in terms of its import</u> I.

(iii) <u>Profits and real wages rise in the North, so much that its consumption of both goods increases</u>.

(iv) <u>In the South, profits rise, but employment, real wages, and consumption of basics all fall</u>.

Proof. We consider first the market-clearing condition in the oil market,

$$X^S_\theta = M^N_\theta \quad . \tag{4.34}$$

From (4.6), (4.9), and (4.6)', this equals

$$\theta^S_S(FT) - \theta^S_D = \theta^N_D \quad . \tag{4.35}$$

From (4.14),

$$\theta_D = b_1 B_S + b_2 I_S \tag{4.36}$$

and from inverting (4.12) and (4.13) we obtain

$$\theta_D = \frac{b_1}{D}(c_2 L - a_2 K) + \frac{b_2}{D}(a_1 K - c_1 L) \quad . \tag{4.37}$$

In view of (4.3) and (4.4), we may rewrite (4.35):

$$\theta_S(FT) + \frac{\alpha}{D}\frac{w}{p_B} M + \frac{\beta r}{D} Q = -\frac{\alpha^N}{D^N}(w/p_B)M^N - \frac{\beta^N r^N}{D^N} Q \tag{4.38}$$

where M and Q are the determinants defined above. Equation (4.38) gives an implicit relation between real wages and the rates of profits in both regions, and the transfer FT, which we denote as

$$\phi\left[r^N, r^S, (w/p_B)^N, (w/p_B)^S, FT\right] = 0 \quad . \tag{4.39}$$

Since factor prices are functions of commodity prices (see (4.27)), we obtain from the substitution of (4.27) into (4.38) a function linking the transfer FT to the prices of B and I:

$$\theta_S(FT) + \frac{\alpha M}{D^2 p_B}(c_2 p^S_B - c_1 p_I + M) + \frac{\beta Q}{D^2}(p_I a_1 - p^S_B a_2 + Q)$$

$$+ \frac{\alpha^N M^N}{(D^N)^2 p^N_B}(c^N_2 p^N_B - c^N_1 p_I + M^N) + \frac{\beta^N}{(D^N)^2} Q(p_I a^N_1 - p_B a_2 + Q) = 0 \quad . \tag{4.40}$$

Equation (4.40) is an implicit function of the form

$$\Gamma(FT, p_I, p_B^N, p_B^S) = 0 \ .$$

However, the prices of basics p_B^S and p_B^N (which may be different since basics are not traded) are themselves functions of the price of industrial goods p_I in equilibrium.

From equation (4.33) we obtain:

$$p_B^N = p_B^N(p_I) \quad \text{and} \quad p_B^S = p_B^S(p_I) \ .$$

Therefore, (4.40) is actually an implicit function of p_I and FT <u>only</u>

$$\Gamma(FT, p_I) = \Gamma(FT, p_I, p_B^S(p_I), p_B^N(p_I)) = 0 \ . \tag{4.41}$$

It is then possible to differentiate implicitly across equilibria and obtain $\partial p_I/\partial FT$, or equivalently its reciprocal

$$\frac{dFT}{dp_I} = -\left[\frac{\partial \Gamma}{\partial p_I}\right] \Big/ \left[\frac{\partial \Gamma}{\partial FT}\right] \ . \tag{4.42}$$

This equation represents the change in the price of industrial goods that follows an increase in the transfer FT. By (4.40) and (4.6)

$$\frac{d\Gamma}{dFT} = \frac{d\theta_S}{dFT} > 0 \ .$$

Therefore the sign of (4.42) is always that of $-\partial \Gamma/\partial p_I$.

We may now compute the derivative $-\partial \Gamma/\partial p_I$. From (4.40) and (4.41) we obtain

$$-\frac{d\Gamma}{dp_I} = -\frac{\partial p_B^S}{\partial p_I}\left(-\frac{\alpha M}{D^2 p_B^2}(M-c_1 p_I) - \frac{a_2 \beta Q}{D^2}\right) + \frac{\alpha c_1 M}{D^2 p_B} - \frac{\beta Q a_1}{D^2}$$

$$-\frac{\partial p_B^N}{\partial p_I}\left(-\frac{\alpha^N M^N}{(D^N)^2 (p_B^N)^2}(M^N - c_1^N p_I) - \frac{a_2^N \beta^N Q^N}{(D^N)^2}\right) + \frac{\alpha^N c_1^N M^N}{(D^N)^2 p_B^N} - \frac{\beta^N Q^N a_1^N}{(D^N)^2} \tag{4.43}$$

From expression (4.43) we may compute the changes in p_I as FT changes, provided we know the signs of the derivatives dp_B^S/dp_I and dp_B^N/dp_I across equilibria.

The next step is therefore to compute the signs of the derivatives of the price of basic goods p_B with respect to the price of industrial goods p_I across equilibria in each region. For this we utilize the expression relating the real wage and the rate of profit in each region, derived from the market-clearing condition $B_S - B_D = 0$:

$$\frac{\alpha c_2}{D} \frac{w}{p_B} - \frac{\beta a_2 r}{D} - \alpha(w/p_B)^2 = 0 \qquad (4.44)$$

(see equation (4.31)), and also the equations relating factor prices to commodity prices:

$$r = \frac{p_I a_1 - p_B a_2 + Q}{D}, \qquad w/p_B = \frac{p_B c_2 - p_I c_1 + M}{Dp_B} \qquad (4.45)$$

(see equation (4.27)). Equation (4.44) is an implicit expression between real wages and profits in each region, denoted $\Lambda(w/p_B, r) = 0$. Since equations (4.44) and (4.45) give real wages and profits as functions of commodity prices, (4.44) actually gives an implicit relation between commodity prices in each region, denoted

$$\psi(p_B, p_I) = \Lambda\left[\frac{w}{p_B}(p_B, p_I), r(p_B, p_I)\right] = 0 \qquad (4.46)$$

From (4.46), by the implicit function theorem, in each region:

$$\frac{dp_B}{dp_I} = -\left[\frac{\partial \psi}{\partial p_I}\right] \Big/ \left[\frac{\partial \psi}{\partial p_B}\right]$$

$$= \left[-\frac{\partial \psi}{\partial (w/p_B)} \frac{\partial (w/p_B)}{\partial p_I} + \frac{\partial \psi}{\partial r} \frac{\partial r}{\partial p_I}\right] \Big/ \left[\frac{\partial \psi}{\partial (w/p_B)} \frac{\partial (w/p_B)}{\partial p_B} + \frac{\partial \psi}{\partial r} \frac{\partial r}{\partial p_B}\right] \qquad (4.47)$$

Furthermore, from (4.45) we find that the partial derivatives

$$\frac{\partial (w/p_B)}{\partial p_I} = -\frac{c_1}{Dp_B} < 0 \qquad (4.48)$$

$$\frac{\partial r}{\partial p_I} = \frac{a_1}{D} > 0 \tag{4.49}$$

$$\frac{\partial (w/p_B)}{\partial p_B} = \frac{p_I c_1 - M}{D p_B^2} > 0 \quad \text{when } p_I > b_2 \tag{4.50}$$

and

$$\frac{\partial r}{\partial p_B} = -\frac{a_2}{D} < 0 \quad . \tag{4.51}$$

We therefore obtain, from (4.47) and (4.51),

$$\frac{dp_B}{dp_I} = \left[\frac{c_1}{Dp_B}\Delta + \frac{a_1 \beta a_2}{D^2}\right] \Big/ \left[\Delta \frac{(p_I c_1 - M)}{Dp_B^2} + \frac{a_2^2 \beta}{D^2}\right] \tag{4.52}$$

where

$$\Delta = \alpha(c_2/d - 2w/p_B) \quad .$$

From relation (4.52) we may now determine the sign of dp_B/dp_I in both the North and the South. First note that dp_B/dp_I is always positive in the North since $p_I > b_2$, so that $p_I c_1 - M > 0$, and $\Delta > 0$ by assumption. In the South $\Delta < 0$, but β is rather small. Therefore, (4.52) is also positive in the South. With this information we may now return to equation (4.40) and compute $\partial r/\partial p_I$. As α is large in the South and β is large in the North, we have from (4.43) that the expression for $-\partial r/\partial p_I$ is dominated by the following terms:

$$\frac{\alpha M}{D^2 p_B^2}(M - c_1 p_I)\frac{dp_B^S}{dp_I} + \frac{\alpha c_1 M}{D^2 p_B} + \frac{dp_B^N}{dp_I}\frac{a_2^N \beta^N Q^N}{(D^N)^2} - \frac{\beta^N Q^N a_1^N}{(D^N)^2} \tag{4.53}$$

Here $M - c_1 p_I = c_1 b_2 - b_1 c_2 - c_1 p_I$ is negative as c_1 is small in the South. Hence the first term is positive (because $M^S < 0$) and dominates the second, which is multiplied by c_1. As $Q^N < 0$, the third term is negative and the fourth positive. But a_2 is small in the North, so that the fourth term dominates. Hence we have that

$$-\frac{\partial \Gamma}{\partial p_I} > 0 .$$

Since $\partial \Gamma/\partial FT = d\theta^S/dFT > 0$, by (4.42) the price of industrial goods p_I rises as the transfer to the South increases, i.e.,

$$dp_I/dFT > 0 . \tag{4.54}$$

We next study the movements of the rate of return in the North r^N as p_I changes. From the national income identity

$$p_I I_D^N = rK - FT .$$

As $I_D^N = I_S^N - X_I^N$ and $p_I X_I^N = X_\theta^S = \theta_D^N$,

$$p_I I_S^N = rK + \theta_D^N - FT .$$

In the North, β is large. We can therefore neglect terms other than those in β, giving

$$p_I = [(-Q/D) + r]/(a_1/D)$$

with

$$\frac{dp_I}{dr} = D/a_1 > 0 . \tag{4.55}$$

Hence as FT rises, p_I rises and the profit rate in the North r^N rises. Knowing how r^N moves enables us to find the sign of the change in the real wage in the North. By (4.31),

$$\frac{ac_2 w}{Dp_B} - \frac{\beta a_2 r}{D} - \alpha(w/p_B)^2 = 0$$

Implicit differentiation gives:

$$\frac{d(w/p_B)}{dr} = \frac{-a_2 \beta}{D\Delta} \tag{4.56}$$

where $\Delta = \alpha(c_2/D - 2w/p_B)$. As $\Delta < 0$ in the North by assumption, we have that

$$\frac{d(w/p_B)}{dr} > 0 \tag{4.57}$$

in the North. Hence an increase in FT raises the real wage in the North, as well as the profit rate. The next step is to show that the consumption levels of B and I rise in the North.

$$I_D^N = rK - FT = \beta r^2 - FT, \qquad \frac{\partial I_D^N}{\partial FT} = 2\beta r \frac{\partial r^N}{\partial FT} - 1 \tag{4.58}$$

which is positive for large β. Also,

$$B_D^N = wL/p_B = \alpha(w/p_B)^2 \tag{4.59}$$

so that B_D^N also rises with FT by (4.57), (4.54), and (4.55). We have now proven point (iii) of Theorem 7.

Next we study the response of trade patterns to FT. By inverting (4.12) and (4.13),

$$X_I^N = I_S^N - I_D^N = \frac{a_1 K}{D} - \frac{c_1 L}{D} - rK + FT .$$

From (4.3) and (4.4)

$$X_I^N = \frac{a_1}{D}\beta r - \frac{c_1 \alpha w}{Dp_B} - \beta r^2 + FT .$$

Hence

$$\frac{dX_I^N}{dr} = \beta\left(\frac{a_1}{D} - 2r\right) - \alpha\frac{c_1}{D}\frac{d(w/p_B)}{dr} + \frac{dFT}{dp_I}\frac{dp_I}{dr} .$$

By the conditions of the theorem, the first term is negative. By (4.57) the second term is negative, and by (4.56) it contains β. As β is large, these terms dominate, and therefore

$$dX_I^N/dr < 0 , \tag{4.60}$$

i.e., the North's exports of the industrial good fall as Ft and hence r^N rise. This implies, of course, that the South's imports of industrial goods fall,

$$dM_I^S/dr^N < 0 \ . \tag{4.61}$$

We next check what happens to the volume of oil traded. This equals oil demanded in the North, θ_D^N, which from (4.23) is

$$-\frac{\alpha w M}{D p_B} - \frac{\beta r Q}{D} \ .$$

Here β is large and Q is negative, by assumption. r rises, by (4.55). Hence

$$\frac{d\theta_D^N}{dFT} = \frac{dX_\theta^S}{dFT} > 0 \ . \tag{4.62}$$

This proves points (i) and (ii) of Theorem 7.

What remains is to study the behavior of the Southern economy. We first show that r^S rises with FT. This is done by showing that $dM_I^S/dr^S < 0$. As

$$\frac{dM_I^S}{dr^S} = \frac{dM_I^S}{dr^N} \frac{dr^N}{dr^S} \ ,$$

this will imply from (4.61) that $dr^N/dr^S > 0$, which in conjunction with (4.54) and (4.55) gives $dr^S/dFT > 0$. From this last inequality the results follow. We therefore compute now the sign of dM_I^S/dr^S.

$$M_I^S = I_D^S - I_S^S = rK + \theta_S + FT - I_S^S = \beta r^2 - \frac{\beta r a_1}{D} + \frac{c_1 \alpha w}{D p_B} + \theta_S + FT$$

$$\frac{dM_I^S}{dr^S} = \beta(2r - a_1/D) + (c_1 \alpha/D)\left(\frac{d(w/p_B)}{dr}\right) + \left(\frac{d\theta_S}{dM_I^S} + \frac{dFT}{dM_I^S}\right)\frac{dM_I^S}{dr^S} \ ,$$

i.e., $\quad \dfrac{dM_I^S}{dr^S}\left(1 - \dfrac{d\theta_S}{dM_I^S} - \dfrac{dFT}{dM_I^S}\right) = \beta(2r - a_1/D) + c_1\alpha/D\left(\dfrac{\partial(w/p_B)}{\partial r}\right) \ .$

Now

$$\frac{d\theta_S}{dM_I^S} = \frac{d\theta_S}{dr^N} \frac{dr^N}{dM_I^S} < 0$$

by (4.61), (4.54), (4.55) and (4.6). Similarly, $dFT/dM_I^S < 0$. By (4.56), $d(w/p_B)/dr < 0$ in the South. As by assumption α^S is large, this establishes that

$$dM_I^S/dr^S < 0 \quad \text{so that} \quad dr^S/dFT > 0. \tag{4.63}$$

It now follows from (4.56) and the fact that $\Delta^S < 0$ by assumption, that real wages in the South fall with FT. It follows immediately from (4.3) and (4.5) that employment and the consumption of basics also fall. This completes the proof of Theorem 7. ∎

Theorem 8. <u>Assume that $M^S > 0$, i.e., $c_1 b_2 - b_1 c_2 > 0$ in the South. Let p_B be small and $p_I > b_2^S$ at the initial equilibrium, with all other conditions as in Theorem 7. Then an increase in the financial transfer FT to the South has the opposite effects to those established in Theorem 7: it leads to a fall in the price of the industrial good p_I and a relative increase in the price of oil p_o, even though oil supplies have increased. The oil exporter's terms of trade therefore improve. In addition, oil exports and the rate of profit in the South decrease. The North exports more industrial goods. Real wages, employment, and consumption of basics increase in the South. In the North, the rate of profit and the real wage decrease.</u>

<u>Proof.</u> As in the proof of Theorem 7, the sign of dFT/dp_I equals that of $-d\Gamma/dp_I$. This is given in equation (4.43), or approximately in (4.53). The latter may also be written as

$$\frac{\alpha M}{D^2 p_B} \left(\frac{dp_B}{dp_I} (M - c_1 p_I) \frac{1}{p_B} + c_1 \right) + \frac{\beta^N Q^N}{(D^N)^2} \left(\frac{dp_B^N}{dp_I} a_2^N - a_1^N \right), \tag{4.64}$$

where parameters and variables are of the South unless otherwise indicated. Now note from (4.52) that for large β^N,

$$dp_B/dp_I \simeq a_1/a_2 \ .$$

Hence the second term in (4.64) is close to zero and (4.64) can be expressed as

$$\frac{\alpha M}{D^2 p_B} \left[\frac{dp_B}{dp_I} \frac{c_1}{p_B} (b_2 - p_I) - \frac{dp_B}{dp_I} \frac{c_2 b_1}{p_B} + c_1 \right]. \qquad (4.65)$$

Under the conditions of Theorem 8 and since $dp_B/dp_I > 0$ by (4.52), (4.65) is negative, proving that the oil exporter's terms of trade improve, i.e., p_I falls with FT.

The rest of the theorem follows immediately. Inequality (4.55) implies that the profit rate in the North falls, and (4.56) implies that real wages in the North fall. Inequality (4.60) tells us that the North's exports (and the South's imports) of industrial goods will increase, and from (4.62) we then know that oil exports of the South fall. Equation (4.63) establishes that the rate of profit in the South falls, and using (4.56) again proves that real wages, employment, and consumption of basic goods all rise in the South. This completes the proof. ■

Theorems 7 and 8 give conditions for opposite effects of a financial transfer FT. The main difference in the conditions of Theorems 7 and 8 are: the sign of the determinant M^S, and the impact that the transfer has on the relative price of industrial goods p_I. The sign of M^S is positive in Theorem 8, and negative in Theorem 7. It seems more plausible that M^S should be negative, since this happens when the basic goods sector in the South uses few capital inputs. Theorem 8 assumes, instead, that the basic goods sector is more capital intensive. The impact of the transfer on prices seems also more plausible in Theorem 7: transfers which are used to develop oil supplies are likely to lead to lower, rather than higher, oil prices. We therefore should expect the relative price of industrial goods to increase. In Theorem 8, the transfer also increases oil supplies, but this leads to higher oil prices, an outcome which appears less plausible.

Clearly, an empirical analysis of the actual conditions is needed to ascertain which set of conditions is applicable. A priori, however, the conditions in Theorem 7 appear more intuitively natural than those in Theorem 8.

C. Conclusions

An inflow of foreign capital, wether used for consumption or investment, inevitably affects the internal equilibrium of the receiving country. Consumption patterns, production patterns, and prices all change. The same is true of the lending country: in changes its consumption pattern by making a loan, and for this reason, and because the equilibrium of its trading partner changes, its own domestic equilibrium alters. A crucial factor in determining these macro effects of a loan is the change in relative prices (oil prices, industrial prices, and prices of basic goods that are not traded). A loan must be one of significant size before having a measurable impact on prices, and for the case of Mexico discussed in Chichilnisky, Heal and McLeod (1984), where the transfer is of the order of 100 billion US dollars, it certainly fits this discription.

It is clear, then, that it is a complex matter to trace the full impacts of a loan from one trading country to another. The model has enabled us to identify these impacts in a rather simple fashion, because of our somewhat stylized assumptions, and to assess the gains and the losses arising from such a loan for different groups within the lending and borrowing countries.

One important feature to emerge is that the loan may have a beneficial effect on the equilibrium of the _lending_ country. This happens when the borrowed funds are used to increase oil supplies, leading to more abundant oil, increased oil exports, and lower oil prices. The terms of trade of the lending country improve, and this leads to higher levels of consumption of both goods in the lending country. Theorem 7 establishes the conditions under which the welfare level in the lending country will rise as a result. In making a social cost-benefit analysis of such a loan, this is a point that should clearly be considered; there is a social return to the loan over and above the rate of interest paid on it. It is possible that even if a major rescheduling that delayed repayment were to happen, the lending country as a whole could nevertheless benefit.

Similar issues apply to the receiving country. The borrowing sector may benefit in commercial terms from the loan, but a social cost-benefit analysis of the loan should also take into account its effects on the overall economic equilibrium.

As Theorem 7 shows, these could be substantially negative. If there has been overspecialization in one sector, thus leading to lower terms of trade for the country, with correspondingly negative welfare effects. In summary, the fact that a loan, if large, may affect the equilibrium pattern of prices and quantities in both countries means that it will have macroeconomic consequences going far beyond its impacts on the profits of the borrowing and lending institutions.

Theorems 7 and 8 have indicated two very different possible outcomes. In one case, the effects are beneficial to the lending and harmful to the borrowing country, while in the other case the opposite is true. The distinguishing feature is the effect of the loan on the oil exporter's terms of trade. In the first case, they worsen, and in the second, they improve. Which of these two outcomes occurs depends on the patterns of factor intensities in the receiving country and the initial price levels. Once these are known, thus establishing whether the loan improves or worsens the receiver's terms of trade, everything else can be traced. Experience indicates that over the last three years, the terms of trade of oil exporters have worsened. While many factors have contributed to this price movement, this suggests that a policy of borrowing to invest in the oil sector might not have been the most favorable to the oil exporter. However, such a policy could be favorable to the lender; it yields more oil at lower prices. Such macro outcomes should be computed when discussing the present situation. The calculus of the debt must go beyond the financial aspects, and must include the macroeconomic effects on prices, imports, and exports of both countries.

It is important to emphasize that we have studied the consequences of granting a loan before this was repaid. The repayments will not have effects that are simply equal and opposite to those of the granting of the loan. The asymmetry arises because, when the loan is made, it is invested or consumed in sectors different than those that will pay the debt. For instance, in this paper the debt was used to build up the production capacity of the oil sector. However, when the load is repaid, this will not, of course, coincide with running down this capacity. Investment is irreversible, and capital stock and machines depreciate. The loan will be repaid by running a balance-of-trade surplus. The effects of running a trade surplus at a constant capacity level in the oil sector are not the opposite of those of running a trade deficit and using the capital inflow to expand oil-producing capacity. As a matter of fact, both could affect the major macro variables in the same direction.

Finally, we point out a connection between the problem that we have studied here and the extensive literature on the transfer problem in international economics. This literature is concerned with the possibility that a transfer of resources from one agent or country to another may benefit the donor and harm the recipient. Chichilnisky (1980) showed for the first time that such an outcome can occur in perfectly competitive markets with a unique and globally stable market equilibrium. This issue has so far been studied only in the context of a barter economy without production in the case of perfectly competitive general equilibrium models. For surveys of these results, see Chichilnisky (1980), Jones (1983), and Geanakoplos and Heal (1983). Our present Theorem 7 provides an example of the transfer paradox in a production economy: resources are transferred from lender to borrower, and the lender gains as a result, even though the receiver expands its production capacity.

References

Chichilnisky, G. (1980), "Basic Goods, Commodity Transfers, and the International Economic Order," J. Development Economics

Chichilnisky, G. (1981a), "Resources and North-South Trade: A Macro Analysis in Open Economies," Working Paper, Columbia University, New York.

Chichilnisky, G. (1981b), "Terms of Trade and Domestic Distribution: Export-Led Growth with Abundant Labor," J. Development Economics

Chichilnisky, G. (1981c), "Oil Prices, Industrial Prices and Outputs: A General Equilibrium Macro Analysis," Working Paper, University of Essex

Chichilnisky, G. (1984), "North-South Trade and Export-Led Growth," J. Development Economics

Chichilnisky, G., G. Heal and D. McLeod (1984), "Resources, Trade and Debt: The Case of Mexico," Working Paper, Columbia University, New York.

Geanakoplos, J., and G. Heal (1983), "A Geometric Explanation of the Transfer Paradox in a Stable Economy," J. Development Economics

Heal, G. and D. McLeod (1983), "Gains from Trade, Stability and Profits", Working Paper, Woodrow Wilson School of Public and International Affairs, Princeton University, New Jersey. J. Development Economics (1984).

Jones, R. (1983), "Notes on the Transfer Paradox in a Three-Agent Setting," Working Paper, University of Rochester, New York. Canad. J. Economics, (1984).

Sterner, T. (1982), "Economic Effects of the Oil Expansion in Mexico," Memo No. 83, University of Gotenberg.

THE ROLE OF MATHEMATICIANS IN NATURAL RESOURCE MODELING

Panel Discussion

R. McKelvey: The general topic for discussion is "The Role of Mathematicians in Natural Resource Modeling." I think there are two ways to interpret that. One is: Is there something special about the training of a mathematician that makes him useful as an individual or as a member of a team, doing resource modeling? That's one kind of question to explore. And another is: what are the interesting problems that would challenge mathematicians?

But I want first to talk briefly about two things that have intrigued me personally about this subject. One is the contribution of Harold Hotelling. The second is the role that has been played by optimal control theory.

In 1931 Harold Hotelling wrote a paper which many people think really started this subject of resource modeling--made it into a coherent science. It was called "The Economics of Exhaustible Resources," and was published in the Journal of Political Economy, April 1931.

The interesting thing is that while JPE is an economics journal, Hotelling was not primarily an economist. His Ph.D. was in mathematics, and shortly after obtaining his Ph.D. he went to work in an applied biology lab at Stanford where he got interested in statistics. Then he went to England and worked for a while with R. A. Fisher, the man who was simultaneously founding a theory of probablistic induction and of genetics. Hotelling became one of the new breed of mathematical statisticians, and indeed a pillar of the Institute of Mathematical Statistics throughout his life.

Hotelling's first papers were in 1925, and that first year he wrote three or four, one of them already on economics, a now recognized contribution to intertemporal utility theory. His next economics paper was the one in 1931, and in the meantime he had published a stream of papers on statistics, mainly multivariant analysis.

If you look at the list of his publications over many years, most of them were in mathematics or statistics journals. Yet he had such an impact on economics that when he was approaching age 65 there were two memorial volumes to him--one by the statisticians and the other by the economists. Hotelling in fact made a number of major contributions to economic theory. He was eventually appointed professor of economics as well as professor of mathematics, and he regularly taught a mathematical-economics course. He was interested in economic theory rather than econometrics. In his long career he worked on applications to biology, medicine, anthropology, and all over the map, but his first love remained mathematical economics.

Hotelling had strong views on applied mathematics. Here is a sample:

> Mathematics, to my way of thinking, is the most general of all subjects. Everything else is more special than mathematics. There is nothing that has a richer profusion of applications, there is nothing that travels over the whole domain of human knowledge as does mathematics. There is no surer key to unlock all sorts of doors than mathematics. Mathematics won't get you all the way, but it will get you into places where you can't enter by any other method. It will supplement all other types of investigation, and help to get out profounder truths than would be possible without its aid. The remarkable thing is that whereas other methods of research are appropriate to special domains, to special kinds of investigations, mathematics--the same, identical mathematics--is applicable to this enormous profusion of different kinds of investigation and is adapted to bringing to light truths of the utmost diversity.
> (Math. Teacher v. 29, 1936)

Back in the 30s Hotelling was arguing that calculus should be a high school subject because, he said, it was important that scientists (including social scientists) should know calculus in order to understand their own fields.

Hotelling's 1931 paper seems to me quite remarkable for its modern flavor. He really anticipated the way people have come to look at resource problems today. The specific issue he was dealing with was the problem of the optimal rate of utilizing the resource of a mine--an exhaustible resource. His was almost exactly the formulation that Geoff Heal showed us yesterday afternoon. Using variational calculus, Hotelling was able to solve the problem rather completely in the case where the costs of extraction are constant. He also probed the more general case where they are variable, depending on the total extraction, the rate of extraction and so on. A bit earlier, there was a paper by Frank Ramsey who modeled the basic problem of capital theory--the problem of savings vs. consumption, the optimal rate of savings. Ramsey used variational calculus, and Hotelling knew of his work.

tal theory and the exhaustion of resources, and recognized that capital theory is inherently dynamic. He anticipated most of the issues that have been studied since in the theory of the mine:

> The static-equilibrium type of economic theory which is now so well developed is plainly inadequate for an industry in which the indefinite maintenance of a steady rate of production is a physical impossibility, and which is therefore bound to decline. How much of the proceeds of a mine should be reckoned as income, and how much as return capital? What is the value of a mine when its contents are supposedly fully known, and what is the effect of uncertainty of estimate? If a mine-owner produces too rapidly, he will depress the price, perhaps to zero. If he produces too slowly, his profits, though larger, may be postponed farther into the future than the rate of interest warrants. Where is his golden mean? And how does this most profitable rate of production vary as exhaustion approaches? Is it more profitable to complete the extraction within a finite time, to extend it indefinitely in such a way that the amount remaining in the mine approaches zero as a limit, or to exploit so slowly that mining operations will not only continue at a diminishing rate forever but leave an amount in the ground which does not approach zero? Suppose the mine is publicly owned. How should exploitation take place for the greatest general good, and how does a course having such an objective compare with that of the profit-seeking entrepreneur? What of the plight of laborers and of subsidiary industries when a mine is exhausted? How can the state, by regulation or taxation, induce the mine-owner to adopt a schedule of production more in harmony with the public good? What about import duties on coal and oil? And for these dynamical systems what becomes of the classic theories of monopoly, duopoly, and free competition?
>
> (JPE v. 39(2))

Hotelling was way ahead of his time, and his work was ignored. First of all resource economics was a backwater subject. In earlier times economists had talked a lot about land as an input factor, but in the 30's the labor theory of value was in dominance. Plus the fact that a lot of what Hotelling wrote was either published in mathematics journals or referred to things published in mathematics journals. For example, he cited G. C. Evans, a well known American mathematician, who had written an introduction to economics for mathematicians, and C.F. Roos, who published in the <u>American Journal of Mathematics</u> and the <u>Transactions of the AMS</u>.

Finally in the sixties, resource economists began looking at these problems. The turn-around was signalled by a volume edited by Mason Gaffney in 1964. Gaffney's volume contained papers that attempted to explain what Hotelling had done all those years before, ironically, re-interpreting him in the language of statics!

Now that mathematicians are taking an active interest in natural resource modeling, it seems fitting to recall this central role played by an applied mathematician at the inception of the subject.

The other thing I want to mention is the special role of optimal control theory, first recognized in the late sixties. I think most mathematicians are aware of the role that theoretical mathematics played in the development of the ideas of quantum mechanics: that John von Neumann had already constructed, in his study of unbounded self-adjoint operators in Hilbert space, the essential basis for formulating quantum theory in a totally natural way.

It seems to me that the development of optimal control theory (for reasons had nothing to do with economics) had somewhat the same kind of impact on resource economics. I want to quote from a 1969 paper by Robert Dorfman, an economist who had been associated with Paul Samuelson and Robert Solow in the early recognition of the importance of linear programming to economics. They were very aware of the role of duality--the notion of the adjoint variable, or shadow price--an idea that also went back to Hotelling, and also to John von Neumann in his competitive game theory.

> "Capital theory," Dorfman said, "is the economics of time. Traditionally, capital theory, like all other branches of economics was studied in the context of stationary equilibrium. Now, the mode of analysis that is confined to a distant ultimate position is poorly suited to the understanding of accumulation and growth. But no other technique seemed available for most of the history of capital theory. For the past fifty years [this was in '69] it has been perceived more or less vaguely that capital theory is formally a problem in the calculus of variations. [And of course that was the way in which Hotelling had conceived it.] But the calculus of variations is regarded as a rather arcane subject by most economists and, besides, in its conventional formulations appears too rigid to be applied to many economic problems. The application of this conceptual tool to capital theory remained peripheral and sporadic until very recently and capital theory remained bound by the very confining limitations of the ultimate equilibrium. All of this has changed abruptly in the past decade

as the result of a revival, or rather reorientation, of the calculus of variations prompted largely by the requirements of space technology. [They were trying to guide rockets to the moon.] In its modern version the calculus of variations is called optimal control theory. It has become deservedly the central tool of capital theory and has given the latter a new lease on life. As a result capital theory has become so profoundly transformed that it has been rechristened growth theory and has come to grips with numerous important practical and theoretical issues that previously could not even be formulated. The main thesis of this paper is that optimal control theory is formerly identical with capital theory and that its main insights can be attained by strictly economic reasoning. This thesis will be supported by deriving the principle theorem of optimal control theory, called the maximum principle, by means of economic analysis."

(Amer. Econ. Rev. v. 59, 1969)

Dorfman then gives what is essentially a heuristic argument--a similar argument is in Colins' book. But the assertion of formal identity is I think fair. The control theory formulation did have a profound affect on the further development of capital theory, because suddenly the mathematics was no longer prohibitively difficult and the mathematical structures permitted direct economic interpretation.

So with that, I would like to open this session for general discussion. First of all, is there anybody on the speakers' platform who wants to make a comment?

G. Heal: I would like to pick up the point you made about Hotelling's role because I think that is interesting. There were, in fact, four or five papers written in the field of mathematical economics in the 1930s which were little understood and little noticed at the time, but which actually provided the basis for a lot of subsequent post-war developments in the area. There were three papers by Hotelling: one on resource depletion, which Bob just quoted; one on the management of public utilities, which was basically a paper about the applications of non-convex analysis in welfare economics; and a paper on imperfect competition. The three papers which Hotelling published on economics in the thirties all were in some sense overlooked at the time but by the 1960s and 1970s gave rise to a very extensive literature. There also were, as Bob mentioned, a couple of papers by F. Ramsey, who was also a statistician, one on optimal growth, and one on taxation, both of which gave rise to a very extensive literature in the 60s and 70s. And there were also a couple of papers by von Neumann. I suppose the first mathematical formalization of the kind of general equilibrium theory that Graciela was talking about earlier was in fact by von Neumann in a paper in the 1930s. He was the first person to realize that you could apply topological techniques to studying the existence and the properties of solutions of general equilibrium systems, and he proved what at that time was a significant advance on existing fixed point theorems in order to establish the existence of solutions to certain types of general equilibrium problems. In addition he started the whole field of game theory. So a lot of the sort of intellectual roots of post-1960 mathematical economics are actually found in work by mathematicians published in the 1930s which worked its way through with a very long time lag, and the main reason for that time lag being that most economists at the time simply couldn't understand what these people were writing about. I find it quite remarkable that they managed to get their work published. I'm delighted that they did, but I think that equivalent people today would have great difficulty in persuading any economics journal to take an article that was as incomprehensible today as Hotelling's article must have been in 1931 or von Neumann's must have been in 1932.

So that's an elaboration of your basic point. Can I go on and make a few other points?

One thing I wanted to remark on is the range of techniques--the range of mathematical techniques--we've heard discussed yesterday and today. We've heard techniques which range from control theory, through functional analysis and optimization, to some form of geometry or differential topology, and a number of other techniques. So one of the things that has been established now is that this area can benefit from application, not just of one or two techniques, but of a wide range of mathematical techniques.

I think that one of the problems that sometimes occurs with fields of applied mathematics is that they get stuck in a particular technical mode. People apply a particular mathematical technique in that area and then for some reason the area gets stuck with that mathematical technique, and the mathematical technique limits what kinds of problems can be posed and what kinds of answers can be generated.

There are certain areas of economics, for example, some mainstream fields of welfare economics where techniques of convex analysis were applied early. All the economists learned these techniques. They didn't learn any other techniques so if they wanted to use mathematics, this was basically the only kind of mathematics they could use. And this fact that there was essentially only one kind of mathematical technique they could use widely, limited very strongly the kinds of problems they could pose and the kinds of questions they could get out.

The area that we're talking about today has benefited from the fact that a wide range of techniques was introduced and therefore we haven't become stuck in any one particular groove. Bob mentioned the massive impact of optimal control theory on capital theory. I'll give another side to that story which is that for five or six years when that was going on, I was the managing editor of one of the main economic theory journals. And everyday of my life for five years, two papers landed on my desk applying optimal control to capital theory! After three or four years there were very rapidly diminishing returns to this and I got extremely bored with it: so I think did most other people. That was an example of the fact that because economists suddenly learned that you could get some mileage out of optimal control theory they put all their graduate students through (to be quite frank) relatively low level courses in optimal control theory, and generations of graduate students came out of the better schools knowing optimal control theory and quite determined to apply it to something.

That generated a whole literature in which people were trying to apply these techniques that they had been taught--the latest and best practice techniques--to whatever problem that could be beaten into a shape which approximated maximization of an integral subject to a differential equation constraint. A lot of technical advances were made in the first five years in that field; in the last five years a lot of paper was wasted.

And that brings me to another point which I want to make on this issue of mathematical education. This is that, as I was saying earlier, we've seen a fair range of techniques used in the literature we've discussed. Anybody with graduate training in mathematics would obviously have no problem at all in following any of the papers that have been discussed. On the other hand we're not actually writing for mathematicians; we're writing for economists. And anyone with a standard graduate training in economics would have immense difficulty in following any of the papers given yesterday or today. And that's a considerable problem for the development of the economics profession.

As I say what tends to happen is that one or two of these techniques get fed through undergraduate courses in economics; students come out with a competent but not outstanding grasp of one or two areas and that circumscribes their subsequent professional development. And in fact it's interesting that it's not only the case that a lot of the intellectual foundations for the works of the 60s and 70s in mathematics of economics were laid by mathematicians back in the 1930s. A lot of the development that then occurred in the 1960s and 1970s was development which was carried out by people who, although they became professional economists, did their Ph.D.'s in mathematics. In fact I know very few people who work in economic theory with Ph.D.'s in economics. Almost all of them have at least some part of their graduate training in mathematics, either pure or applied.

So a problem the economics profession faces is that the mathematical development of the subject requires a level of technical training which most economics graduate schools simply don't provide. As a result there is in this area a market for people who have a much broader range of technical training of the type that comes out of mathematics degree courses or mathematics graduate courses.

G. Chichilnisky: I am very pleased about Bob McKelvey's introduction on Harold Hotelling. Hotelling used to work at my Department at Columbia, where he wrote those seminal articles mentioned before. He was the thesis adviser of Kenneth Arrow, with whom I worked at Harvard between '74 and '78, and who has a Nobel Prize in economics. Kenneth Arrow, who is now at Stanford, has tremendous admiration and respect for Hotelling, and was one of the very very few students that Hotelling had at Columbia. Hotelling left Columbia and went to Duke University to join their statistics department, apparently dissatisfied with his life at Columbia.

Hotelling produced extremely original work that the economics profession took a very long time to recognize. As was mentioned before, some of the most fundamental work in economics has been carried out by people that were trained as mathematicians, and an inordinate number of Nobel Prize winners in economics are people who have the Ph.D. in mathematics and not in economics. There is a very close connection between the fields of economics and mathematics. And this connection does not go in one direction only: It is not true that advanced economics is produced by applying tricks in some field in mathematics. Economics helps the production and development of new mathematics as well.

For example, fix point theorems, were invented almost simultaneously within economics and mathematics; other examples are programming and optimization techniques, a number of algorithms for computing fixed points, and new results in nonlinear analysis and in algebraic topology. All of these tools were produced by people who were investigating economic problems with a mathematical mind. It is not desirable to get a mathematical technique in your hands and then stop thinking, and apply the technique until you die. Instead we have to <u>produce</u> mathematical theorems needed to solve the economics problems.

There is no distance between economics and mathematics. The questions are different, but mathematics is in the eyes of the beholder. When looking at an economic question the mathematician will find a mathematical question, and will hopefully find also mathematical answers. To tell an economist, "this is the mathematics you should know and apply" is a mistake. The reverse problem is that of a mathematician who has learned a neat technique and tries to apply it without much regard to the problem itself.

Mathematics in the last few years has been more involved in solving problems than in formulating new and important problems. When a discipline, be that mathematics or anything else, spends too much time in solving older problems and too little time formulating new problems, it is preparing itself for its death. Economics could be, among other disciplines, an important source of new mathematical problems. I mean new topology, new algebraic geometry, new number theory, should be developed to solve our problem in economics, much the same way that applied physics led Poincare to initiate and develop algebraic topology.

The problems themselves should guide us in the choice of the mathematics. We cannot make a division between the problem and the techniques. Neither should we stop thinking about a problem because it does not recur to lead to "important" mathematics. I think that the only way to solve a problem is to let the problem dominate us completely. The problem is what matters. We want to solve the problem. Whatever it takes! Whether this fits or not with the current mathematical vogue is irrelevant. To look at the technique rather than at the problem may stop us from producing important mathematics--from advancing in mathematics. As already mentioned Poincare developed beautiful topology for solving an extremely applied problem. And many many years later we are still trying to answer questions that he posed.

C. Clark: I certainly think those are very profound remarks. And the only point I would like to develop further is that we do have specilizations in the world. Graciela and I and all the rest at this table think of ourselves as specialists in mathematics, and I found in my own work that it's incredibly valuable for me to keep in contact with specialists in those other areas, not only economics but in the resources field--the people who know about the resources. I spend a lot of time visiting biological stations in America and overseas and talking to fisheries biologists. I traveled around the world on my sabbatical and visited fishery and biology stations all over the place. And I've worked with people in those stations as well as with economists. And I think that's one point that neither of the previous speakers alluded to specifically. Perhaps it was a point where Hotelling himself fell down; I don't know to what extent he actually communicated or worked with other people that were experts in those fields. We really have in the modern world a society of specialists in science.

But it is true, and this is one place (Bob McKelvey frequently points this out) that mathematicians really have a unique role that they can play here; namely that they sort of speak the universal language. Most mathematicians I know have a humility, (and if they don't, they should develop a humility) in facing the real world and in facing these people and in trying to understand their problems. I would like to underline the importance of this: if you're going to do important mathematical research on important problems, make sure that you are studying the important problems. And to be sure that you're studying the important problems, talk with the people who are not professional mathematicians who have those problems.

Pat Kenschaft (Montclair State College): A couple weeks ago I was told I'm going to teach a modeling course at Montclair State for the first time. It was taught once before but that person has vanished. I thought you might be willing or able to give me a little advice because I know some of the people here are doing this kind of thing...now, or soon will be doing it. What can be done for undergraduates--what should we be trying to do? I love your statements--all of them. They sound so good. But I try to think what I can do with juniors and seniors who are reasonably able at mathematics. And I wonder how much should I try to bite off in one semester and how do I go about doing it?

R. Plant: We've talked about it before, but maybe I could elaborate a little bit. I think one thing that everybody else has been saying, but let me rephrase it in a slightly different way, is that for most people who do this kind of work--their education doesn't end, and in fact it begins--after their formal education ends. That people who get stuck in a rut do that because their training was pretty largely confined to solving problems that were assigned to them. I think a modeling course especially is a perfect opportunity to give students the opportunity to be creative and to go out on their own with a sort of vaguely defined problem or assignment and just see what they can do with it. To study the literature, read the literature, see where previous modeling efforts have broken down, try to criticize these modeling efforts and see how they might do better.

I think that one aspect of university education, and I think it's especially true of mathematics education, that is neglected, is the training or emphasis of creativity. And the ability to learn on your own instead of just being spoon fed. And I think a modeling course is a good opportunity to do this.

P. Kenschaft: I love the idea. And I would like to give it a whirl. I have been told repeatedly at this conference and at other places that it has never been done at the undergraduate level with any modicum of success. I've been told of a lot of unsuccessful attempts in the last couple of days and elsewhere. Can anybody give me any positive encouragement?

R. Plant: We certainly have done it at Davis.

P. Kenschaft: At the undergraduate level?

R. Plant: Yes.

C. Clark: There is a book--the author's name is Bender. There are a lot of books on modeling, but most just present finished models.

P. Kenschaft: One can certainly talk about how you create models and follow one of the textbooks...

G. Chichilnisky: May I make a suggestion? Perhaps what you could do is pick three textbooks: One for applications of models to social sciences or economics; another one for applications of models to biology, and the third would be for applications to physics. And then tell the students they can choose modeling in one of these three applied areas. Then they can learn about different applications, make expositions about these models, and how to improve them. It seems desirable to give them a choice of applications--biological, physical and social sciences.

P. Kenschaft: And actually have them do their own models?

G. Chichilnisky: No, that may be too much.

R. Plant: Right, what they have to do is to look at the models in a semester or a quarter. To understand their own modeling, they would have to first learn the subject and then construct the model. And that is a lot to expect. What they can do is look at other people's models, learn them on their own through self study rather than your giving a lecture, understand them, understand what their weak spots are, (which, as Colin mentioned in his talk, is something that's sorely neglected so often in our training), and try to use their creativity to indicate how they might improve them.

G. Chichilnisky: One way you can go about doing that is the following: If people find it difficult at some point to conceptualize significant variations on a model, then you can ask them to take a model which exists and (if it is simple enough--let's say a system of two or three equations) simulate it; simulate the solutions. And then tell them, "See what happens when you change either one of the equations or one of the parameters of the model with the solutions." If they know basic programming, they can do that. And I suspect that your students will know basic programming, so they should be able to do that.

G. Heal: One of the difficulties one always faces in applying mathematics is: given a problem--a real world problem--to find the mathematical structure in that problem. And that's something which most courses don't teach. And if you're taking a course on homology theory or whatever it is and you're given a problem set, then you know for a start that these are problems in homology; it doesn't take much imagination to know which book you're supposed to be looking in for solutions. If a mathematician is given a problem in economics, it's not clear exactly whether this is a problem in homology, in optimization or functional analysis or whatever. So one of the difficult first steps in any mathematical modeling exercise is: given a problem, to decide what kind of mathematics is best used to present this. The kind of course that you're talking about I think is the kind of medium through which one could develop some feeling for that issue among students. Now I taught a modeling course once and I don't want to claim that it was tremendously successful. I found it a difficult course to teach. But one thing I did that was similar to what was being suggested was to take a number of relatively well-known and very successful models in different areas and have the students look at what happens if you change the assumptions in those models. And what you find in a lot of areas is that there are particular combinations of assumptions about how you represent the system, which fit together neatly and give you nice results. And other combinations of assumptions--you may change just one of those assumptions-- which somehow fit together much less well making it much much harder to get analytical solutions.

One of the skills that anybody working in modeling in a particular area learns rapidly, but learns kind of intuitively, is which assumptions, which combinations of assumptions go together to give you something manageable and tractible, and give a nice structure to the problem. And which combinations, although they might look at least as attractive, somehow just don't work out well. And you can learn that sort of thing by just taking an existing model, which is well documented, and going through and replacing some of the simplifying assumptions by other assumptions which look equally simple, but which you will find then will make the model behave quite differently or make it behave very badly.

I did that in particular with a number of economic models which were typically characterized by simple assumptions about consumer behavior, simple assumptions about producer behavior, some simple assumptions about market structure. And if you work in this game, you rapidly learn that there are some combinations which work out nicely, and give you clean results, while some other combinations and other permutations don't fit together at all to give anything decent.

It might be useful for the students in a couple of fields to try doing that kind of thing. Then they'll begin to realize, through a learning by doing process, why people make the particular simplifying assumptions that they do make. And then I think for an outsider coming in to look at a mathematical model, which is always very very simple anyway relative to the true problem, it's useful for them to get some feeling for why those particular simplifying assumptions were made. You don't usually get that feeling until you get right inside the problem and work out why it doesn't work out any other way.

C. Clark: We have an undergraduate modeling course and I've always shied away from teaching it because I think it would be tremendously difficult to really do a good job. What I would like is for somebody to collect a list-- not very long, 10 or 20--simple unstructured modeling problems that undergraduates with minimal training in any field could tackle. I have one that I put on the floor, and it's this: How often should you sharpen a saw? Now there must be thousands of problems like that. Just simple problems like that that could come from physics or operations research or whatever and if somebody could make a collection of those...so what I'm suggesting, Bob, is that we get this yellow sheet and pass it around. Everybody must have one or two problems like this in the back of his/her mind--no matter how silly. Just write it down and maybe we could get at least fifty problems.

R. McKelvey: Very well. Let's try it.

Charles Pace (Idaho State): I'm glad you started off with Hotelling. But there are other economists of some stature who have been somewhat more reserved in their endorsement of mathematics and economics. I think of Keynes' preface to the French edition of The General Theory where he says "Economic theory is a bundle of ideas which the author throws to the reader, and if the reader catches it, the reader then is not bogged down by a lot of details."

Later on in The General Theory he makes it pretty clear that that attack is directed at the application of mathematics in economics.

G. Chichilnisky: Did you know that Keynes was a mathematician?

C. Pace: That's true; he also loved ballet.

G. Heal: And ballerinas?

C. Pace: Nicolas Georgeson-Roegen talks about arithomania in his Entropy Law in Economics, the practice in economics of applying arithmetic operations when you have no reason for doing so or no understanding. Other people--Koopmans, who also used mathematics extensively, talks about how successful application of a technique in one field doesn't create any presumption in favor or against application in another field. He goes on to say that the danger is one of erecting too many analogies and forcing. Now that doesn't seem to be inconsistent with what several of the panelists have mentioned but at the same time, I'd like to get a clearer idea of the limitations. Everybody is saying mathematics should not serve as the guide--but at the same time I don't think anybody would say, that mathematicians are just janitors--that they're just going to perform this sort of custodial service. So maybe you could sort of trace out the area between that--the idea of mathematics not serving as a guide but at the same time mathematicians serving as more than custodians.

C. Clark: I myself don't see anything wrong with mathematics. I don't know what you mean by custodian, but I'm not embarrassed if I work for several years on a problem and don't discover any actually new mathematical theories or methods or techniques or theorems or anything. As Geoff said, few economists, I would add biologists and so on, have the range of techniques that a person with graduate training in mathematics has. Most of us--even if we at the moment don't know any stochastics, don't know any probability theory and so on; we can learn that if we have to, and apply that. Much more easily certainly than say the biologist who may have taken one course in statistical methods or something like that. So it's nice that we will definitely encounter new mathematical problems that will excite us, eventually. But personally, I don't take that as something I derive my choice of problems by. I look at a problem; I find a problem interesting; I attack it in whatever way I can; and why should we be worried about it--is that what you mean by "custodial service to the other science"? We're the custodians of mathematics. The other people have their own difficulties.

C. Pace: No, by "custodial" I just mean clean up the inconsistencies. In other words, does mathematics have something more to offer than just cleaning up the inconsistencies, and where is the difference between what it has to offer and serving as the guide? I mean we want to strike an intermediate position between mathematics, as say the iron mike, versus mathematics as the janitor. But I'm not sure how you people--iron mike is where you set your compass and it steers the ship. Where is the intermediate position there?

R. McKelvey: I think I understand a little better what you're saying. For example, in quantum mechanics, in the 1970s I think, the mathematicians finally cleaned up the model of the helium atom, which the physicists had lost interest in half a century before. Now that may be a legitimate and worthwhile role but I think we've been talking here about whether mathematicians have a role at the cutting edge of developing the science. My impression is that for most mathematicians the answer is probably no. Because most mathematicians are not really comfortable out in the inductive world, in the chaos of science at the frontier. Because most mathematicians temperamentally prefer to work in structures that are well set--where the foundations are firm and where they can use mainly deductive approaches. There's a difference between science and mathematics--it seems to me. The fundamental difference is that mathematics is a deductive subject and science is not exclusively that, although deduction plays a role in it.

But for any mathematician who is willing to move out from deductive mathematics into empirical science, the rewards can be very great. And that is because of the special training and insights that he brings with him. It isn't just mathematicians who can do this. I think of Robert May, for example, a physicist who moved into population biology.

Those people now are fairly rare. Anyone who is willing to cross a frontier has to be very brave or possibly very arrogant. But it seems to me we ought to be cultivating the frontiers between the sciences; we should be trying to find people and persuade them that it's fruitful to work in those areas. It's often the most fruitful kind of thing to do because no single tradition has all the threads in hand.

Now the viewpoint as to what is applied mathematics and how applied mathematicians should perform, has been evolving in the last decade or fifteen years in what I think is a very good way. It used to be that most applied mathematicians regarded their role to be to go out with a bundle of techniques and find applications that would fit them. That's not a very flattering statement, but I think it had been true for many people. But the viewpoint now, among applied mathematicians, is quite different. It is the idea that applied mathematicians ought to be able to work on a lot of different subjects for which they have no specialized training. They ought to be willing to go out and try. And what they bring to that endeavor is not a particular set of tools, but rather the ability to penetrate the necessary literature in the field and also in mathematics. Colin Clark probably knew very little about control theory when he got into this business. He discovered that that's what he needed to learn. And he learned what he needed to learn in order to do what he wanted to do. That's the attitude applied mathematicians should have, I think, toward what their role is.

But it's important that they not confuse themselves about <u>who</u> they are; an applied mathematician is not usually also an economist or a physicist, or a chemist, biologist, or whatever. And he doesn't have all the insights those people have. Unless he's willing to spend a lot of time and energy to also become a scientist of a particular sort, then it's incumbent upon him to do what Colin was emphasizing, namely, to find appropriate collaborators or at least talk to people. It seems to me that a lot of applied mathematicians today are doing exactly that. They are going out and seeking collaborators. People I know work with half a dozen different collaborators, sometimes in totally different fields. I think that's a very healthy way to proceed. I would like to see us training our students to do that.

<u>G. Chichilnisky</u>: I don't think mathematics is a deductive science. When you do mathematics you search for something you think may be there. You search for relationships; you prove new relationships. In fact relationships don't exist until you prove them. And that's not a deductive process. You don't deduce particulars from generals or anything like that--you're going into new relationships, new structures. In fact you invent your subject.

You can invent exactly what you want to study--what you want to find--what you think is important--you can define new structures. What's interesting about mathematics is how often invented structures that came from somebody's mind have tremendously useful applications. Even though when they were invented, they only seemed to be theoretically powerful, or beautiful. I think that we should not separate mathematics from any other pursuit of science. The applied mathematician should not view himself as an amputated economist or biologist, etc. Everybody should seek interaction and understanding of problems with the practitioners, like Colin Clark said. That's crucial, because that's where interesting issues come from, and we should be pressured by reality into thinking of important relationships. There is nothing more powerful than reality. There is nothing more interesting than reality.

If you view yourself as an amputated biologist, an amputated economist, psychologically you're not prepared to take the chance of studying a new applied problem. And I think we need to develop the new mathematics which is needed. And I think we need in mathematics more people prepared to take chances. The safe attitude of--let's clean up what somebody else has done; let's see if we can present this in better mathematical vocabulary; let's prove a little bit more general theorem--has proven to be unsuccessful and frustrating both in mathematics and in economics. The fifties were a time where you had a blooming of ideas in economics. Since then, very talented people, very well trained, have devoted themselves to clean-up what was understood to be interesting ideas of somebody else. I observe a similar trend in mathematics. It seems a pity.

It was generally believed that mathematical economics consists of formalizing somebody else's ideas, and the ideas come from somewhere else. Now we are showing that this is not true. It took thirty years and a lot of disagreement. Still many people will tell you that the role of the mathematical economist is to formalize what the informal economist says--that the beautiful idea come from elsewhere that the mathematical economists role into try to put it in a formulas. And I think that's wrong. To approach any problem with an amputated framework of mind is defeatist.

R. McKelvey: I hope I wasn't saying exactly those things that you're criticizing! I was arguing that a mathematician needs to find collaborators unless he's prepared to learn everything that the collaborators would know. But I'm not arguing that he should wait until twenty years after those people have finished the work. I want him to be involved in the process of discovery. So far as the notion of mathematics as a deductive science, I wasn't talking about the creative process but rather the process of verification of truth in mathematics, which is by deduction--not by going out and looking in nature. I think that modeling is a very different kind of activity from that.